Modern Data Mining Algorithms in C++ and CUDA C

Recent Developments in Feature Extraction and Selection Algorithms for Data Science

Timothy Masters

Apress®

Modern Data Mining Algorithms in C++ and CUDA C: Recent Developments in Feature Extraction and Selection Algorithms for Data Science

Timothy Masters
Ithaca, NY, USA

ISBN-13 (pbk): 978-1-4842-5987-0   ISBN-13 (electronic): 978-1-4842-5988-7
https://doi.org/10.1007/978-1-4842-5988-7

Managing Director, Apress Media LLC: Welmoed Spahr
Acquisitions Editor: Steve Anglin
Development Editor: Matthew Moodie
Coordinating Editor: Mark Powers

Cover designed by eStudioCalamar

Cover image by Krissia Cruz on Unsplash (www.unsplash.com)

Distributed to the book trade worldwide by Apress Media, LLC, 1 New York Plaza, New York, NY 10004, U.S.A. Phone 1-800-SPRINGER, fax (201) 348-4505, e-mail orders-ny@springer-sbm.com, or visit www.springeronline.com. Apress Media, LLC is a California LLC and the sole member (owner) is Springer Science + Business Media Finance Inc (SSBM Finance Inc). SSBM Finance Inc is a **Delaware** corporation.

For information on translations, please e-mail editorial@apress.com; for reprint, paperback, or audio rights, please email bookpermissions@springernature.com.

Apress titles may be purchased in bulk for academic, corporate, or promotional use. eBook versions and licenses are also available for most titles. For more information, reference our Print and eBook Bulk Sales web page at http://www.apress.com/bulk-sales.

Any source code or other supplementary material referenced by the author in this book is available to readers on GitHub via the book's product page, located at www.apress.com/9781484259870. For more detailed information, please visit http://www.apress.com/source-code.

Printed on acid-free paper

Table of Contents

About the Author

Timothy Masters received a PhD in mathematical statistics with a specialization in numerical computing. Since then, he has continuously worked as an independent consultant for government and industry. His early research involved automated feature detection in high-altitude photographs while he developed applications for flood and drought prediction, detection of hidden missile silos, and identification of threatening military vehicles. Later he worked with medical researchers in the development of computer algorithms for distinguishing between benign and malignant cells in needle biopsies. For the last 20 years, he has focused primarily on methods for evaluating automated financial market trading systems. He has authored twelve books on practical applications of predictive modeling:

Practical Neural Network Recipes in C++ (Academic Press, 1993)

Signal and Image Processing with Neural Networks (Wiley, 1994)

Advanced Algorithms for Neural Networks (Wiley, 1995)

Neural, Novel, and Hybrid Algorithms for Time Series Prediction (Wiley, 1995)

Assessing and Improving Prediction and Classification (Apress, 2018)

Deep Belief Nets in C++ and CUDA C: Volume I: Restricted Boltzmann Machines and Supervised Feedforward Networks (Apress, 2018)

Deep Belief Nets in C++ and CUDA C: Volume II: Autoencoding in the Complex Domain (Apress, 2018)

Deep Belief Nets in C++ and CUDA C: Volume III: Convolutional Nets (Apress, 2018)

Data Mining Algorithms in C++ (Apress, 2018)

Testing and Tuning Market Trading Systems (Apress, 2018)

Statistically Sound Indicators for Financial Market Prediction: Algorithms in C++ (KDP, 2019, 2nd Ed 2020)

Permutation and Randomization Tests for Trading System Development: Algorithms in C++ (KDP, 2020)

About the Technical Reviewer

Michael Thomas has worked in software development for more than 20 years as an individual contributor, team lead, program manager, and vice president of engineering. Michael has more than 10 years of experience working with mobile devices. His current focus is in the medical sector, using mobile devices to accelerate information transfer between patients and healthcare providers.

CHAPTER 1

Introduction

Serious data miners are often faced with thousands of candidate features for their prediction or classification application, with most of the features being of little or no value. Worse still, many of these features may be useful only in combination with certain other features while being practically worthless alone or in combination with most others. Some features may have enormous predictive power, but only within a small, specialized area of the feature space. The problems that plague modern data miners are endless.

My own work over the last 20+ years in the financial market domain has involved examination of decades of price data from hundreds or thousands of markets, searching for exploitable patterns of price movement. I could not have done it without having a large and up-to-date collection of data mining tools.

Over my career, I have accumulated many such tools, and I still keep up to date with scientific journals, exploring the latest algorithms as they appear. In my recent book *Data Mining Algorithms in C++*, published by the Apress division of Springer, I presented many of my favorite algorithms. I now continue this presentation, divulging details and source code of key modules in my *VarScreen* variable screening program that has been made available for free download and frequently updated over the last few years. The topics included in this text are as follows:

> **Hidden Markov models** are chosen and optimized according to their multivariate correlation with a target. The idea is that observed variables are used to deduce the current state of a hidden Markov model, and then this state information is used to estimate the value of an unobservable target variable. This use of memory in a time series discourages whipsawing of decisions and enhances information usage.

> **Forward Selection Component Analysis** uses forward and optional backward refinement of maximum-variance-capture components from a subset of a large group of variables.

© Timothy Masters 2020
T. Masters, *Modern Data Mining Algorithms in C++ and CUDA C*,
https://doi.org/10.1007/978-1-4842-5988-7_1

This hybrid combination of principal components analysis with stepwise selection lets us whittle down enormous feature sets, retaining only those variables that are most important.

Local Feature Selection identifies predictors that are optimal in localized areas of the feature space but may not be globally optimal. Such predictors can be effectively used by nonlinear models but are neglected by many other feature selection algorithms that require global predictive power. Thus, this algorithm can detect vital features that are missed by other feature selection algorithms.

Stepwise selection of predictive features is enhanced in three important ways. First, instead of keeping a single optimal subset of candidates at each step, this algorithm keeps a large collection of high-quality subsets and performs a more exhaustive search of combinations of predictors that have joint but not individual power. Second, cross-validation is used to select features, rather than using the traditional in-sample performance. This provides an excellent means of complexity control, resulting in greatly improved out-of-sample performance. Third, a Monte-Carlo permutation test is applied at each addition step, assessing the probability that a good-looking feature set may not be good at all, but rather just lucky in its attainment of a lofty performance criterion.

Nominal-to-ordinal conversion lets us take a potentially valuable nominal variable (a category or class membership) that is unsuitable for input to a prediction model and assign to each category a sensible numeric value that can be used as a model input.

As is usual in my books, all of these algorithm presentations begin with an intuitive overview, continue with essential mathematics, and culminate in complete, heavily commented source code. In most cases, one or more sample applications are shown to illustrate the technique.

The *VarScreen* program, available as a free download from my website TimothyMasters.info, implements all of these algorithms and many more. This program can serve as an effective variable screening program for data mining applications.

CHAPTER 2

Forward Selection Component Analysis

The algorithms presented in this chapter are greatly inspired by the paper "Forward Selection Component Analysis: Algorithms and Applications" by Luca Puggini and Sean McLoone, published in *IEEE Transactions on Pattern Analysis and Machine Intelligence*, December 2017, and widely available for free download on various Internet sites (just search). However, I have made several small modifications that I believe make it somewhat more practical for real-life applications. The code for a subroutine that performs all of these algorithms is in the file FSCA.CPP.

Introduction to Forward Selection Component Analysis

The technique of principal components analysis has been used for centuries (or so it seems!) to distill the information (variance) contained in a large number of variables down into a smaller, more manageable set of new variables called *components* or *principal components*. Sometimes the researcher is interested only in the nature of the linear combinations of the original variables that provide new component variables having the property of capturing the maximum possible amount of the total variance inherent in the original set of variables. In other words, principal components analysis can sometimes be viewed as an application of descriptive statistics. Other times the researcher wants to go one step further, computing the principal components and employing them as predictors in a modeling application.

However, with the advent of extremely large datasets, several shortcomings of traditional principal components analysis have become seriously problematic. The root cause of these problems is that traditional principal components analysis computes the new variables as linear combinations of *all* of the original variables. If you have been presented with thousands of variables, there can be issues with using all of them.

One possible issue is the cost of obtaining all of these variables going forward. Maybe the research budget allowed for collecting a huge dataset for initial study, but the division manager would look askance at such a massive endeavor on an ongoing basis. It would be a lot better if, after an initial analysis, you could request updated samples from only a much smaller subset of the original variable set.

Another issue is interpretation. Being able to come up with descriptive names for the new variables (even if the "name" is a paragraph long!) is always good, something we often want to do (or *must* do to satisfy critics). It's hard enough putting a name to a linear combination of a dozen or two variables; try understanding and explaining the nature of a linear combination of two thousand variables! So if you could identify a much smaller subset of the original set, such that this subset encapsulates the majority of the independent variation inherent in the original set, and then compute the new component variables from this smaller set, you are in a far better position to understand and name the components and explain to the rest of the world what these new variables represent.

Yet another issue with traditional principal components when applied to an enormous dataset is the all too common situation of groups of variables having large mutual correlation. For example, in the analysis of financial markets for automated trading systems, we may measure many families of market behavior: trends, departures from trends, volatility, and so forth. We may have many hundreds of such indicators, and among them we may have several dozen different measures of volatility, all of which are highly correlated. When we apply traditional principal components analysis to such correlated groups, an unfortunate effect of the correlation is to cause the weights within each correlated set to be evenly dispersed among the correlated variables in the set. So, for example, suppose we have a set of 30 measures of volatility that are highly correlated. Even if volatility is an important source of variation (potentially useful information) across the dataset (market history), the computed weights for each of these variables will be small, each measure garnering a small amount of the total "importance" represented by the group as a whole. As a result, we may examine the weights, see nothing but tiny weights for the volatility measures, and erroneously conclude that volatility does not carry much importance. When there are many such groups, and especially if they do not fall into obvious families, intelligent interpretation becomes nearly impossible.

The algorithms presented here go a long way toward solving all of these problems. They work by first finding the single variable that does the best job of "explaining" the total variability (considering all original variables) observed in the dataset. Roughly speaking, we say that a variable does a good job of explaining the total variability if knowledge of the value of that variable tells us a lot about the values of all of the other variables in the original dataset. So the best variable is the one that lets us predict the values of all other variables with maximum accuracy.

Once we have the best single variable, we consider the remaining variables and find the one that, *in conjunction with the one we already have*, does the best job of predicting all other variables. Then we find a third, and a fourth, and so on, choosing each newly selected variable so as to maximize the explained variance when the selected variable is used in conjunction with all previously selected variables. Application of this simple algorithm gives us an ordered set of variables selected from the huge original set, beginning with the most important and henceforth with decreasing but always optimal importance (conditional on prior selections).

It is well known that a greedy algorithm such as the strictly forward selection just described can produce a suboptimal final set of variables. It is always optimal in a certain sense, but only in the sense of being conditional on prior selections. It can (and often does) happen that when some new variable is selected, a previously selected variable suddenly loses a good deal of its importance, yet nevertheless remains in the selected set.

For this reason, the algorithms here optionally allow for continual refinement of the set of selected variables by regularly testing previously selected variables to see if they should be removed and replaced with some other candidate. Unfortunately, we lose the ordering-of-importance property that we have with strict forward selection, but we gain a more optimal final subset of variables. Of course, even with backward refinement we can still end up with a set of variables that is inferior to what could be obtained by testing every possible subset. However, the combinatoric explosion that results from anything but a very small universe of variables makes exhaustive testing impossible. So in practice, backward refinement is pretty much the best we can do. And in most cases, its final set of variables is very good.

There is an interesting relationship between the choice of whether to refine and the number of variables selected, which we may want to specify in advance. This is one of those issues that may seem obvious but that can still cause some confusion. With strict forward selection, the "best" variables, those selected early in the process, remain the same, regardless of the number of variables the user decides to keep. If you decree in advance that you want a final set to contain 5 variables but later decide that you want 25, the first 5 in that set of 25 will be the same 5 that you originally obtained. There is a certain intuitive comfort in this property. But if you had specified that backward refinement accompany forward selection, the first 5 variables in your final set of 25 might be completely different from the first 5 that you originally obtained. This is a concrete example of why, with backward refinement, you should pay little attention to the order in which variables appear in your final set.

Regardless of which choice we make, if we compute values for the new component variables as linear combinations of the selected variables, it's nice if we can do so in such a way that they have two desirable properties:

1) They are standardized, meaning that they have a mean of zero and a standard deviation of one.

2) They are uncorrelated with one another, a property that enhances stability for most modeling algorithms.

If we enforce the second property, orthogonality of the components, then our algorithm, which seeks to continually increase explained variance by its selection of variables, has a nice attribute: the number of new variables, the components, equals the number of selected variables. The requirement that the components be orthogonal means that we cannot have more components than we have original variables, and the "maximize explained variance" selection rule means that as long as there are still variables in the original set that are not collinear, we can compute a new component every time we select another variable. Thus, all mathematical and program development will compute as many components as there are selected variables. You don't have to use them all, but they are there for you if you wish to use them.

The Mathematics and Code Examples

In the development that follows, we will use the notation shown in the following text, which largely but not entirely duplicates the notation employed in the paper cited at the beginning of this chapter. Because the primary purpose of this book is to present practical implementation information, readers who desire detailed mathematical derivations and proofs should see the cited paper. The following variables will appear most often:

> $\mathbf{X}$ – The m by v matrix of original data; each row is a case and each column is a variable. We will assume that the variables have been standardized to have mean zero and standard deviation one. The paper makes only the zero-mean assumption, but also assuming unit standard deviation simplifies many subsequent calculations and is no limitation in practice.

$\mathbf{x}_i$ – Column i of $\mathbf{X}$, a column vector m long.

m – The number of cases; the number of rows in $\mathbf{X}$.

v – The number of variables; the number of columns in $\mathbf{X}$.

k – The number of variables selected from $\mathbf{X}$, with $k \leq v$.

$\mathbf{Z}$ – The m by k matrix of the columns (variables) selected from $\mathbf{X}$.

$\mathbf{z}_i$ – Column i of $\mathbf{Z}$, a column vector m long.

$\mathbf{S}$ – The m by k matrix of new "component" variables that we will compute as a linear transformation of the selected variables $\mathbf{Z}$. The columns of $\mathbf{S}$ are computed in such a way that they are orthogonal with zero mean and unit standard deviation.

$\mathbf{s}_i$ – Column i of $\mathbf{S}$, a column vector m long.

The fundamental idea behind any principal components algorithm is that we want to be able to optimally approximate the original data matrix $\mathbf{X}$, which has many columns (variables), by using the values in a components matrix $\mathbf{S}$ that has fewer columns than $\mathbf{X}$. This approximation is a simple linear transformation, as shown in Equation (2.1), in which Θ is a k by v transformation matrix.

$$\hat{\mathbf{X}} = \mathbf{S}\Theta \tag{2.1}$$

In practice, we rarely need to compute Θ for presentation to the user; we only need to know that, by the nature of our algorithm, such a matrix exists and plays a role in internal calculations.

We can express the error in our approximation as the sum of squared differences between the original $\mathbf{X}$ matrix and our approximation given by Equation (2.1). In Equation (2.2), the double-bar operator just means to square each individual term and sum the squares. An equivalent way to judge the quality of the approximation is to compute the fraction of the total variance in $\mathbf{X}$ that is explained by knowing $\mathbf{S}$. This is shown in Equation (2.3).

$$Err = \left\| \hat{\mathbf{X}} - \mathbf{X} \right\|^2 \tag{2.2}$$

$$Explained\ Variance = 1 - \frac{\left\| \hat{\mathbf{X}} - \mathbf{X} \right\|^2}{\left\| \mathbf{X} \right\|^2} \tag{2.3}$$

With these things in mind, we see that we need to intelligently select a set of k columns from **X** to give us **Z**, from which we can compute **S**, the orthogonal and standardized component matrix. Assume that we have a way to compute, for any given **S**, the minimum possible value of Equation (2.2) or, equivalently, the maximum possible value of Equation (2.3), the explained variance. Then we are home free: for any specified subset of the columns of **X**, we can compute the degree to which the entire **X** can be approximated by knowledge of the selected subset of its columns. This gives us a way to "score" any trial subset. All we need to do is construct that subset in such a way that the approximation error is minimized. We shall do so by forward selection, with optional backward refinement, to continually increase the explained variance.

Maximizing the Explained Variance

In this section, we explore what is perhaps the most fundamental part of the overall algorithm: choosing the best variable to add to whatever we already have in the subset that we are building. Suppose we have a collection of variables in the subset already, or we are just starting out and we have none at all yet. In either case, our immediate task is to compute a score for each variable in the universe that is not yet in the subset and choose the variable that has the highest score.

The most obvious way to do this for any variable is to temporarily include it in the subset, compute the corresponding component variables **S** (a subject that will be discussed in a forthcoming section), use Equation (2.1) to compute the least-squares weights for constructing an approximation to the original variables, and finally use Equation (2.3) to compute the fraction of the original variance that is explained by this approximation. Then choose whichever variable maximizes the explained variance.

As happens regularly throughout life, the obvious approach is not the best approach. The procedure just presented works fine and has the bonus of giving us a numerical value for the explained variance. Unfortunately, it is agonizingly slow to compute. There is a better way, still slow but a whole lot better than the definitional method.

The method about to be shown computes a different criterion for testing each candidate variable, one with no ready interpretation. However, the authors of the paper cited at the beginning of this chapter present a relatively complex proof that the rank order of this criterion for competing variables is the same as the rank order of the explained variance for the competing variables. Thus, if we choose the variable that has the maximum value of this alternative criterion, we know that this is the same variable that would have been chosen had we maximized the explained variance.

We need to define a new notation. We have already defined $\mathbf{Z}$ as the currently selected columns (variables) in the full dataset $\mathbf{X}$. For the purposes of this discussion, understand that if we have not yet chosen any variables, $\mathbf{Z}$ will be a null matrix, a matrix with no columns. We now define $\mathbf{Z}_{(i)}$ to be $\mathbf{Z}$ with column i of $\mathbf{X}$ appended. Of course, if $\mathbf{Z}$ already has one or more columns, it is assumed that column i does not duplicate a column of $\mathbf{X}$ already in $\mathbf{Z}$. We are doing a trial of a new variable, one that is not already in the set. Define $\mathbf{q}_{j(i)}$ as shown in Equation (2.4), recalling that $\mathbf{x}_j$ is column j of $\mathbf{X}$. Then our alternative criterion for trial column (variable) i is given by Equation (2.5).

$$q_{j(i)} = \mathbf{Z}_{(i)}^{\mathrm{T}}\mathbf{x}_j \tag{2.4}$$

$$Crit_i = \sum_{j=1}^{v}\left[q_{j(i)}^{\mathrm{T}}\left(\mathbf{Z}_{(i)}^{\mathrm{T}}\mathbf{Z}_{(i)}\right)^{-1} q_{j(i)} \right] \tag{2.5}$$

Let's take a short break to make sure we are clear on the dimensions of these quantities. The "universe of variables" matrix $\mathbf{X}$ has m rows (cases) and v columns (variables). So the summation indexed by j in Equation (2.5) is across the entire universe of variables. The subset matrix $\mathbf{Z}_{(i)}$ also has m rows, and it has k columns, with $k=1$ when we are selecting the first variable, 2 for the second, and so forth. Thus, the intermediate term $\mathbf{q}_{j(i)}$ is a column vector k long. The matrix product that we are inverting is k by k, and each term in the summation is a scalar.

It's an interesting and instructive exercise to consider Equation (2.5) when $k=1$, our search for the first variable. I'll leave the (relatively easy) details to the reader; just remember that the columns of $\mathbf{X}$ (and hence $\mathbf{Z}$) have been standardized to have zero mean and unit standard deviation. It should require little work to verify that each term in the summation of Equation (2.5) is the squared correlation between variable i and variable j. So we are selecting as our first variable whichever one has the greatest average correlation with all of the other variables. It seems reasonable. On a side note, it is nice to subtract one from this criterion (because the sum includes its correlation with itself) and divide by $v-1$ to get the actual mean correlation and then print this in a table for the user's edification. I do this in the code that will appear later, and I suggest that you do the same, as it's a nice little touch.

Code for the Variance Maximization Criterion

This section presents a subroutine that evaluates the alternative criterion just described, enabling selection of the best variable to add to the subset by forward selection. Complete source code is in the file FSCA.CPP.

But first, we need to do a little preparatory work. In the process of evaluating the alternative criterion, we will need to compute dot products of pairs of variables many, many times. It would be wasteful to keep computing the same quantity over and over. Thus, we compute the matrix of all possible paired dot products of **X** columns in advance and save them in a matrix that I call covar. It deserves this name because I divide each dot product by m, the number of cases. The columns of **X** have been centered to have zero mean, so this is a covariance matrix. In fact, because the columns have also been scaled for unit standard deviation, covar is also a correlation matrix. But know that this scaling is my own choice to make reported weights commensurate; the selection algorithm does not require unit standard deviation, though it does require centering to zero mean.

The following code fragment computes covar from the n_cases by npred **X** matrix in x. We need to compute only the lower triangle, as the matrix is symmetric. Also, there is no need to compute the diagonal, because standardization guarantees that the diagonals are all 1.0. If you choose to avoid standardization, you'll need to compute the diagonals.

```
for (i=1 ; i<npred ; i++) {         // Initialize for summing
  for (j=0 ; j<i ; j++)
    covar[i*npred+j] = 0.0 ;
  }

for (i=0 ; i<n_cases ; i++) {       // Sum across all cases (rows of x)
  xptr = x + i * npred ;            // Point to this case
  for (j=1 ; j<npred ; j++) {
    dtemp = xptr[j] ;
    for (k=0 ; k<j ; k++)
      covar[j*npred+k] += dtemp * xptr[k] ;  // Cumulate each dot product
    }
  }
```

```
for (j=1 ; j<npred ; j++) {          // Convert dot products to covariances
   for (k=0 ; k<j ; k++) {
      covar[j*npred+k] /= n_cases ;
      covar[k*npred+j] = covar[j*npred+k] ; // The matrix is symmetric
      }
   }

for (j=0 ; j<npred ; j++)            // No need to compute diagonal; we know it
   covar[j*npred+j] = 1.0 ;
```

Okay, the preliminary work is done. We now tackle the subroutine that takes as its input the identities of the variables (columns in X) already in the subset of kept variables (which will be empty for the first variable), as well as the identity of a trial variable to append to the kept subset. It then computes and returns the alternative criterion associated with this expanded kept subset. So to do a forward step, we just call this subroutine for each variable not yet kept and choose whichever variable produces the greatest criterion. The calling parameter list is as follows:

```
double newvar_crit (
   int npred ,             // Number of predictors (columns) in predictor matrix x
   double *covar ,         // Covariance/correlation matrix (scaled x'x)
   int nkept ,             // Number of columns of x kept so far
   int *kept_columns ,     // The columns kept so far
   int trial_col ,         // Trial column for appending to kept columns
   double *work2 ,         // Work array ncomp * ncomp
   double *work3 ,         // Work array ncomp * ncomp
   double *work4 ,         // Work array ncomp * ncomp + 2 * ncomp
   double *work5 ,         // Work array ncomp
   int *work6              // Work array ncomp
   )
```

In this code, the variable npred plays the role of v in the mathematics presented in the prior section. The size specifications for the work arrays are in terms of ncomp, which we have not seen before. Technically, we need only nkept+1 instead of ncomp, but ncomp is the maximum possible size of the kept subset, a parameter specified by the user. So in the calling program, we allocate memory according to the maximum possible subset size and we are safe thereafter.

The first step is to fetch the **Z'Z** matrix from the covariance matrix that we computed in advance and place it in the work array work2. Because we are augmenting the existing subset of nkept variables by a trial variable, for the purposes of the criterion evaluation we now have nkept+1 variables. The following loop first fetches the variables already in the kept subset and then appends the trial variable as the last row and column in the trial matrix.

Finally, we invert **Z'Z**. In the calling program, which we will see later, we take precautions to ensure that the full covariance matrix is nonsingular, which in turn guarantees that all submatrices are nonsingular. Still, it's good form to provide an out in case the matrix is singular (indicated by a nonzero return value).

```
{
  int i, j, k, irow, new_kept, ret_val ;
  double sum, crit, dtemp ;

  new_kept = nkept + 1 ;

/*
Fetch the Z'Z matrix from covar matrix (actually correlations), then invert it.
*/

  for (i=0 ; i<new_kept ; i++) {
    if (i < nkept)
      irow = kept_columns[i] ;
    else
      irow = trial_col ;
    for (j=0 ; j<nkept ; j++)
      work2[i*new_kept+j] = covar[irow*npred+kept_columns[j]] ;
    work2[i*new_kept+nkept] = covar[irow*npred+trial_col] ;
    }

  ret_val = invert ( new_kept , work2 , work3 , &dtemp , work4 , work6 ) ;
  if (ret_val) // Should never happen
    return -1.e60 ;
```

At this point, work3 contains the inverse of **Z'Z**, and of course this inverse is also symmetric. We can now go on to compute the criterion.

The outermost loop (over *j*) performs the summation of Equation (2.5). The first inner loop (over *i*) pulls the *q* vector of Equation (2.4) from the covar matrix. It should now be clear why we computed covar in advance for reuse!

```
crit = 0.0 ;
for (j=0 ; j<npred ; j++) {     // Sum Equation 22 in the paper
  for (i=0 ; i<nkept ; i++)     // Fetch q into work5.
    work5[i] = covar[j*npred+kept_columns[i]] ;
  work5[nkept] = covar[j*npred+trial_col] ;
```

We now compute the outer product terms for this j, doing the diagonal first and then doing the symmetric off-diagonal terms. We have to do the computation for only the lower triangle, because the matrix is symmetric. So we compute for just this triangle and then double it as we add it into the criterion that we are summing.

```
  sum = 0.0 ;
  for (i=0 ; i<new_kept ; i++)     // Diagonal
    sum += work5[i] * work5[i] * work3[i*new_kept+i] ;
  crit += sum ;

  sum = 0.0 ;
  for (i=1 ; i<new_kept ; i++) {    // Sum for lower triangle
    dtemp = work5[i] ;
    for (k=0 ; k<i ; k++)
      sum += dtemp * work5[k] * work3[i*new_kept+k] ;
  }
  crit += 2.0 * sum ; // Matrix is symmetric so account for both sides
  } // For j

return crit ;
}
```

Some readers might be wondering about the fact that to compute covar, we divided by the number of cases to make it a covariance/correlation matrix. Equation (2.5) calls for the actual $\mathbf{Z'Z}$ matrix. But close examination of this equation reveals that any scaling washes out, having no effect on the computed criterion.

Backward Refinement

Backward refinement (removing a currently kept variable and replacing it with a different one) involves nothing more than a variation on the forward selection routine just seen. Even though there is much code duplication, I'll show the complete subroutine here in order to make sure its operation is clear. The calling parameter list is as follows:

```
double substvar_crit (
   int npred ,              // Number of predictors (columns) in predictor matrix x
   double *covar ,          // Covariance matrix (scaled x'x)
   int nkept ,              // Number of columns of x kept so far
   int *kept_columns ,      // The columns (index in covar) kept so far
   int old_col ,            // Trial column (index in kept_columns) to be replaced
   int new_col ,            // Column (index in covar) that replaces old_col
   double *work2 ,          // Work array ncomp * ncomp
   double *work3 ,          // Work array ncomp * ncomp
   double *work4 ,          // Work array ncomp * ncomp + 2 * ncomp
   double *work5 ,          // Work array ncomp
   int *work6               // Work array ncomp
   )
```

The only point of possible confusion in these parameters is the nature of old_col and new_col. The first, old_col, refers to the variable that is to be potentially replaced, and hence it is an index into the list of currently kept columns, kept_columns. The second, new_col, refers to the replacement candidate, a variable not already kept that we may want to substitute for old_col. It is an index into covar, the matrix of pairwise dot products.

Like we did for forward selection, we fetch the **Z'Z** matrix from covar. But first we preserve the status quo before making the substitution.

```
saved_col = kept_columns[old_col] ;   // Save it so we can restore it when done
kept_columns[old_col] = new_col ;

for (i=0 ; i<nkept ; i++) {
   irow = kept_columns[i] ;
   for (j=0 ; j<nkept ; j++)
      work2[i*nkept+j] = covar[irow*npred+kept_columns[j]] ;
   }
```

We invert **Z'Z**, making provision for singularity even though we are guaranteed that it will never be singular.

```
ret_val = invert ( nkept , work2 , work3 , &det , work4 , work6 ) ;
if (ret_val) // Should never happen
    return -1.e60 ;
```

The code now is practically identical to what we saw for forward selection, except that because we are not appending a trial variable, everything is a little cleaner. See the previous section if any of this code is not clear. The final step is to restore the variable that was present originally.

```
crit = 0.0 ;
for (j=0 ; j<npred ; j++) {        // Sum Equation (2.5)
    for (i=0 ; i<nkept ; i++)      // Fetch q into work5.
        work5[i] = covar[j*npred+kept_columns[i]] ;

    // Compute outer product term for this j, diagonal first, then symmetric off-diagonal
    sum = 0.0 ;
    for (i=0 ; i<nkept ; i++)
        sum += work5[i] * work5[i] * work3[i*nkept+i] ;
    crit += sum ;

    sum = 0.0 ;
    for (i=1 ; i<nkept ; i++) {
        dtemp = work5[i] ;
        for (k=0 ; k<i ; k++)
            sum += dtemp * work5[k] * work3[i*nkept+k] ; // This line is the time eater!
    }
    crit += 2.0 * sum ; // Matrix is symmetric so account for both sides
} // For j

kept_columns[old_col] = saved_col ; // Restore the original
return crit ;
}
```

In order to use this subroutine for backward refinement, it would need to be inside a double loop. The outer loop tries each of the currently kept variables, and the inner loop tries substituting every variable in the universe **X** that is not currently kept. If the inner loop finds a substitution that improves the criterion, we perform that substitution.

The subroutine SPBR (*Single Pass Backward Refinement*) implements that very algorithm, calling substvar_crit() inside a double loop. Its calling parameter list is as follows:

```
int SPBR (
  int npred ,             // Number of predictors (columns) in predictor matrix x
  int *preds ,            // Indices into database of user-specified predictors
  double *covar ,         // Covariance matrix (x'x)
  int nkept ,             // Number of columns of x kept so far
  int *kept_columns ,     // The columns (index in covar) kept so far, modified if better
  double *work2 ,         // Work array ncomp * ncomp
  double *work3 ,         // Work array ncomp * ncomp
  double *work4 ,         // Work array ncomp * ncomp + 2 * ncomp
  double *work5 ,         // Work array ncomp
  int *work6 ,            // Work array ncomp
  double *best_crit       // Returns final best criterion
  )
```

Those calling parameters should be familiar by now, so we'll skip directly to the inner workings of this routine. As a baseline reference, we need to find the criterion for the current subset of variables. Of course we've already computed it recently, so if we were desperate, we could save that value to avoid recomputing it now. But that would add complexity to the algorithm, and compared to the work involved in the upcoming doubly nested loop, recomputing it once here adds only insignificant computation time. So I choose simplicity.

```
*best_crit = substvar_crit ( npred , covar , nkept , kept_columns , 0 , kept_columns[0] ,
              work2 , work3 , work4 , work5 , work6 ) ;
```

The outer loop passes through all currently kept variables; old_col specifies the column in kept_columns that we are testing for replacement. For each old_col we compute the criterion that would be obtained by substituting each variable (new_col) for the current variable old_col. Of course, it makes sense to do this only for variables that are not already in kept_columns. We certainly don't want the same variable to appear twice!

There is one subtlety that we can use to avoid some wasted effort. When this routine is called, it is guaranteed (by my program design) that the last entry in kept_columns is the variable added or evaluated most recently.

Thus, when we get to the last column, if no changes have been made, there is no point in doing the last column. We know it's the best, as will be discussed on Page 33.

```
refined = 0 ;              // Flag that no substitutions have been made yet.
for (old_col=0 ; old_col<nkept ; old_col++) {   // Consider replacing each variable

  // If we are at the last variable and no replacements were done, we are finished.
  if (old_col == nkept-1 && ! refined)
    break ;

  // Test every variable that is not currently kept
  best_col = -1 ; // Will flag if we found a better variable for old_col
  for (new_col=0 ; new_col<npred ; new_col++) {   // Trial replacement for old_col

    // Skip this variable 'new_col' if it is already kept
    for (i=0 ; i<nkept ; i++) {
      if (new_col == kept_columns[i])   // Is this trial variable already kept in subset?
        break ;
    }
    if (i < nkept)                       // True if already kept, so skip it
      continue ;

    // Temporarily substitute variable 'new_col' into kept 'old_col' and compute criterion
    crit = substvar_crit ( npred , covar , nkept , kept_columns , old_col , new_col ,
                  work2 , work3 , work4 , work5 , work6 ) ;

    if (crit > *best_crit) {       // Did we improve?
      *best_crit = crit ;
      best_col = new_col ;
    }
  } // For new_col (testing non-kept variables for replacing that in old_col)

  if (best_col >= 0) { // Did we find one that is a good substitute for variable 'old_col'?
    kept_columns[old_col] = best_col ; // Perform the replacement
    refined = 1 ;
  }
} // For old_col (checking columns for replacement)

return refined ;
}
```

Multithreading Backward Refinement

Backward refinement is by far the dominant time eater in the complete algorithm. That should be obvious because we have an outer loop of all currently kept variables, and a loop inside that of all not-currently-kept variables, and inside that (in substvar_crit()) a loop through all variables in the universe, and inside that a doubly nested loop that evaluates the terms of Equation (2.5). Wow. That's loop nesting five deep! In fact, in the listing for substvar_crit() shown starting on Page 14, there is a single line in the code that accounts for the vast majority of the total compute time. Thus, if we are going to multithread any part of the code to take advantage of modern multicore processors, backward refinement is the place to do it. That is the subject of this section.

As a quick aside, as long as we are talking about compute time, I should note an important issue that may occur to some readers. Matrix inversion is a stereotypically slow operation, and we have to invert the $\mathbf{Z'Z}$ matrix repeatedly whenever we evaluate the criterion, whether doing forward or backward operations. The authors of the paper cited at the beginning of this chapter present an elegant algorithm for avoiding repeated inversion, instead building up the inverse one step at a time as variables are added or substituted. However, I performed extensive runtime profiling of this application using very large datasets, as much as 2000 variables and 6000 cases. The absolute greatest fraction of runtime devoted to matrix inversion that I ever encountered in any test was less than one-tenth of one percent of the total runtime! Thus I see no point in doing anything to speed up the matrix inversion; it would be pointless effort.

Back to multithreading. By the way, readers who are not interested in multithreading their code, a significant complexification, may safely skip this section. It just demonstrates how the SPBR() routine shown in the prior section can be enhanced to spread out the computation to separate threads that compute in parallel, thus greatly speeding operation.

As is nearly always the case, instead of passing a mass of parameters to the threaded routine (which Windows generally disallows), we copy all required parameters into a data structure and then pass that structure to the threaded routine via a wrapper routine that will be launched as multiple threads. Here is the data structure we use here, along with the wrapper routine:

```
typedef struct {
   int npred ;              // Number of predictors (columns) in predictor matrix x
   double *covar ;          // Covariance matrix (x'x)
   int nkept ;              // Number of columns of x kept so far
```

```
    int *kept_columns ;    // The columns (index in covar) kept so far, modified if better
    int old_col ;          // Trial column (index in kept_columns) to be replaced
    int new_col ;          // Column (index in covar) that replaces old_col
    double *work2 ;        // Work array ncomp * ncomp
    double *work3 ;        // Work array ncomp * ncomp
    double *work4 ;        // Work array ncomp * ncomp + 2 * ncomp
    double *work5 ;        // Work array ncomp
    int *work6 ;           // Work array ncomp
    double crit ;          // Computed criterion returned here
} FSCA_PARAMS ;

static unsigned int __stdcall substvar_threaded ( LPVOID dp )
{
    ((FSCA_PARAMS *) dp)->crit = substvar_crit (
            ((FSCA_PARAMS *) dp)->npred ,
            ((FSCA_PARAMS *) dp)->covar ,
            ((FSCA_PARAMS *) dp)->nkept ,
            ((FSCA_PARAMS *) dp)->kept_columns ,
            ((FSCA_PARAMS *) dp)->old_col ,
            ((FSCA_PARAMS *) dp)->new_col ,
            ((FSCA_PARAMS *) dp)->work2 ,
            ((FSCA_PARAMS *) dp)->work3 ,
            ((FSCA_PARAMS *) dp)->work4 ,
            ((FSCA_PARAMS *) dp)->work5 ,
            ((FSCA_PARAMS *) dp)->work6 ) ;
    return 0 ;
}
```

There is a critical issue involved here. *Every thread must have its own private set of work areas (work2–work6) and kept_columns.* Thus, when these arrays are allocated, we must allocate the required quantity *times the maximum number of threads that will be launched.* We'll soon see how to keep these areas separate for each thread.

Here is the calling parameter list, identical to that for the nonthreaded version. We also see that for each possible thread, we allocate an FSCA_PARAMS area as well as a thread handle.

```
static int SPBR_threaded (
   int npred ,              // Number of predictors (columns) in predictor matrix x
   int *preds ,             // Indices into database of user-specified predictors
   double *covar ,          // Covariance matrix (x'x)
   int nkept ,              // Number of columns of x kept so far
   int *kept_columns ,      // The columns (index in covar) kept so far, modified if better
   double *work2 ,          // Work array ncomp * ncomp
   double *work3 ,          // Work array ncomp * ncomp
   double *work4 ,          // Work array ncomp * ncomp + 2 * ncomp
   double *work5 ,          // Work array ncomp
   int *work6 ,             // Work array ncomp
   double *best_crit        // Returns final best criterion
   )
{
   int i, k, old_col, new_col, best_col, ithread, n_threads, max_threads ;
   int empty_slot, refined, ret_val ;
   double crit ;
   FSCA_PARAMS params[MAX_THREADS] ;
   HANDLE threads[MAX_THREADS] ;
```

We set the number of threads that we will use here according to the user parameter max_threads_limit. Like we did in the nonthreaded version, we evaluate the criterion with the current subset of kept variables. Then we set the parameters that will be the same for each thread as well as assign separate, private work areas and kept_columns arrays for each thread.

```
   max_threads = max_threads_limit ;

   *best_crit = substvar_crit ( npred , covar , nkept , kept_columns , 0 , kept_columns[0] ,
                     work2 , work3 , work4 , work5 , work6 ) ;

   for (ithread=0 ; ithread<max_threads ; ithread++) {
     params[ithread].npred = npred ;
     params[ithread].covar = covar ;
     params[ithread].nkept = nkept ;
     params[ithread].kept_columns = kept_columns + ithread * nkept ;
```

```
for (i=0 ; i<nkept ; i++)
   params[ithread].kept_columns[i] = kept_columns[i] ;
params[ithread].work2 = work2 + ithread * nkept * nkept ;
params[ithread].work3 = work3 + ithread * nkept * nkept ;
params[ithread].work4 = work4 + ithread * (nkept * nkept + 2 * nkept) ;
params[ithread].work5 = work5 + ithread * nkept ;
params[ithread].work6 = work6 + ithread * nkept ;
} // For ithread
```

In that code block, observe that each thread gets its own copy of kept_columns, and the contents of this array are copied to each thread's private array. This is necessary because substvar_crit(), as you may recall, temporarily changes the contents of this array. Also observe how each thread gets its own independent set of work arrays. Be sure you allocate this memory accordingly!

The outer loop in this threaded version is identical to the outer loop in the nonthreaded version (SPBR()), except that we also have to copy the index of the column up for replacement into the parameter array for each thread. See the discussion of SPBR() if this code block is not clear.

```
refined = 0 ;          // Flag that no substiututions have been made yet.
for (old_col=0 ; old_col<nkept ; old_col++) { // Consider replacing each variable
   for (i=0 ; i<max_threads ; i++)
      params[i].old_col = old_col ;

   // If we are now to the last variable and no replacements were done, we are finished.
   if (old_col == nkept-1 && ! refined)
      break ;
   best_col = -1 ; // Will flag if we found a better v ariable for old_col
```

Here's where things start to get messy. For each old_col, we will multithread the trial replacement variables. Initialize that no threads are running and all thread slots are empty.

```
n_threads = 0 ;              // Counts threads that are active
for (i=0 ; i<max_threads ; i++)
   threads[i] = NULL ;

empty_slot = -1 ; // After full, will identify the thread that just completed
```

Also initialize the first new_col to be the first variable in the universe that is not already in the kept subset.

```
// Start at the first new_col not already kept
new_col = 0 ;   // Index (in covar) of trial replacement predictor
while (new_col < npred) {
  for (i=0 ; i<nkept ; i++) {
    if (new_col == kept_columns[i])    // Is this new_col already in the kept set?
      break ;
    }
  if (i == nkept)      // True if new_col is not already kept
    break ;
  ++new_col ;       // Already kept so advance to the next in the universe
  }
```

We use an "endless" loop to start threads and wait for threads. There are several code blocks within this loop. The first checks for the user pressing the ESCape key. If so, we compact all executing threads into a contiguous group at the start of the thread array and wait for them all to finish. Then we close all of their handles, post a message to the user acknowledging that they interrupted the program, and return to the caller.

```
   for (;;) {   // ------> Main thread loop processes all trial candidates
/*
   Handle user ESCape
*/
     if (escape_key_pressed || user_pressed_escape ()) {
       for (i=0, k=0 ; i<max_threads ; i++) {
         if (threads[i] != NULL)
           threads[k++] = threads[i] ;
         }
       ret_val = WaitForMultipleObjects ( k , threads , TRUE , 50000 ) ;
       for (i=0 ; i<k ; i++)
         CloseHandle ( threads[i] ) ;
       audit ( "ERROR: User pressed ESCape during FSCA" ) ;
       return 0 ;
       }
```

The next code block in the endless loop is the thread launcher, which executes only if the next trial replacement column new_col is still in our universe of contenders. We initialized empty_slot to be –1, where it stays while we are filling the thread slots (executing threads). After all threads are running and one finishes, its slot, now available for reuse, is assigned to empty_slot. In either case, we let k be the thread slot that we will now launch. Its only parameter that must be set at this point is new_col, the trial replacement variable. We launch the thread and make provision for a rare but possible launch error.

```
if (new_col < npred) {      // If there are still some to do
  if (empty_slot < 0)       // Negative while we are initially filling the queue
    k = n_threads ;
  else
    k = empty_slot ;
  params[k].new_col = new_col ;
  threads[k] = (HANDLE) _beginthreadex ( NULL , 0 , substvar_threaded ,
                                &param s[k] , 0 , NULL ) ;
  if (threads[k] == NULL) {
    // tell user about this rare but terrible problem
    for (i=0 ; i<n_threads ; i++) {      // Close all running threads
      if (threads[i] != NULL)
        CloseHandle ( threads[i] ) ;
    }
    return 0 ;
  }

  ++n_threads ;
  ++new_col ;
  // Skip this variable 'new_col' if it is already kept
  while (new_col < npred) {
    for (i=0 ; i<nkept ; i++) {
      if (new_col == kept_columns[i])
        break ;
    }
```

```
    if (i == nkept) // True if new_col is not already kept
      break ;
    ++new_col ;
    }
  } // if (new_col < npred)
```

After the thread is launched, the launching routine returns immediately. We increment n_threads, the counter of running threads, and we advance new_col to the next variable that is not already in the kept subset.

At some point in this "endless" loop, we have to see if there are no longer any threads running and exit the loop if this is the case. This is as good a place as any.

```
if (n_threads == 0)   // Are we done?
  break ;
```

The next code block handles the full suite of threads running, and there are more threads to add as soon as some are done. We sit here and wait for a thread to finish. The wait routine returns the index of the thread that just completed. The wrapper routine listed on Page 19 placed the computed criterion in the crit member of the parameter structure, so we fetch it and see if we set a new record. If so, we update the record and record the variable that produced it. Finally, we make this thread's slot available for reuse (empty_slot), close the thread, set its handle pointer to NULL, and decrement the counter of running threads.

```
if (n_threads == max_threads && new_col < npred) {
  ret_val = WaitForMultipleObjects ( n_threads , threads , FALSE , 500000 ) ;
  if (ret_val == WAIT_TIMEOUT || ret_val == WAIT_FAILED ||
    ret_val < 0 || ret_val >= n_threads) {
    // Inform the user of this rare but serious error
    return 0 ;
    }

  crit = params[ret_val].crit ;
  if (crit > *best_crit) {
    *best_crit = crit ;
    best_col = params[ret_val].new_col ;
    }
```

```
empty_slot = ret_val ;
CloseHandle ( threads[empty_slot] ) ;
threads[empty_slot] = NULL ;
--n_threads ;
}
```

The final code block in the loop handles the situation that all required work has been started and now we are just waiting for all threads to finish. We dutifully prepare for the rare possibility of the wait routine failing. It returns when all threads are complete. We pass through all threads, seeing if their criteria set any new records and keeping track of the best column. We must also close the thread handle to free up system resources.

```
else if (new_col >= npred) {
  ret_val = WaitForMultipleObjects ( n_threads , threads , TRUE , 500000 ) ;
  if (ret_val == WAIT_TIMEOUT || ret_val == WAIT_FAILED ||
    ret_val < 0 || ret_val >= n_threads) { // Inform user of this rare, serious error
    return 0 ;
  }

  for (i=0 ; i<n_threads ; i++) {
    crit = params[i].crit ;
    if (crit > *best_crit) {
      *best_crit = crit ;
      best_col = params[i].new_col ;
    }
    CloseHandle ( threads[i] ) ;
  }
  break ; // We are done
  }
} // Endless loop which threads computation of criterion for all trial new_col
```

That's about it. We initialized best_col to –1, so if it is now nonnegative, we update the master copy of kept_columns, as well as all thread copies, and flag that refinement took place.

```
if (best_col >= 0) { // Did we find a variable that is a good substitute for old_col?
  kept_columns[old_col] = best_col ;
  for (ithread=0 ; ithread<max_threads ; ithread++) {
```

```
      for (i=0 ; i<nkept ; i++)
        params[ithread].kept_columns[i] = kept_columns[i] ;
      }
    refined = 1 ;
    }
  } // For old_col (checking columns for replacement)

  return refined ;
}
```

Orthogonalizing Ordered Components

If we perform strictly forward selection, never doing backward refinement, the selected variables are ordered in importance, starting with the most important and continuing in decreasing importance, conditional on knowing the values of all prior variables. Even if we destroy order by backward refinement, we can still order the selected variables within the subset. In many applications, this ordering is a useful piece of information.

We could just use the raw values of the selected variables for our application, whether it be as predictors for a model or any other purpose. But we established earlier (and it is generally well known) that there are advantages to having the newly computed component variables, which are derived as linear combinations of the original variables, be orthogonal. Among other things, this implies that they are completely uncorrelated, a property that helps many model training systems. It also means that there is no redundancy (at least in linear terms) in the information contained in the variables. This can sometimes make interpretation of their contributions easier.

So we need a way to compute the new component variables from the selected subset in such a way that they are orthogonal but the ordering of importance (variance capture from the complete universe of variables) is preserved.

Luckily, there is an easy way to do this: Gram–Schmidt orthogonalization. We will actually use the well-known modification of this procedure which is mathematically equivalent to the original but has improved stability for numerical computation in which small floating-point errors are possible. We will also standardize the components to zero mean and unit standard deviation.

The procedure works as follows: define the first component variable as the first selected variable but scaled to unit length (not unit standard deviation yet). To get the second component, copy the second selected variable. Find its projection on the first component and subtract that projection from it and then standardize it to unit length. To get the third component, copy the third selected variable. Subtract its projection on the first component and then subtract its projection on the second component. Because the first two components are orthogonal, this second subtraction will not damage the orthogonality that we created with the first subtraction. Normalize this third component to unit length. Continue like this: for each new component, copy its corresponding ordered variable and then subtract its projection on each of the already computed components and normalize to unit length. When we have finished all components, we rescale them to unit standard deviation by multiplying by the square root of the number of cases. This final standardization is not necessary, but it makes them into people-friendly numbers, not tiny little numbers that might not even print well in many programs.

Here is the simple calling parameter list. It is legal to let the input and output matrices be the same array, in which case the output overwrites the input.

```
int GramSchmidt (      // Normally returns 0, returns 1 if not full ncols rank
    int nrows ,        // Number of rows in in/out matrices
    int ncols ,        // And number of columns
    double *input ,    // Input matrix, not changed
    double *output     // Output of orthonormalized matrix; may be input to overwrite
    )
```

The first step is to copy the first input column to the first output column and normalize it to unit length.

```
sum = 0.0 ;

for (irow=0 ; irow<nrows ; irow++) { // Do the first column
    dtemp = input[irow*ncols] ;
    output[irow*ncols] = dtemp ;
    sum += dtemp * dtemp ;
    }

sum = sqrt ( sum ) ;
if (sum == 0.0)
    return 1 ;

for (irow=0 ; irow<nrows ; irow++)
    output[irow*ncols] /= sum ;
```

We now process the remaining columns. For each new column icol, we first copy it from the input to the output. The inner loop makes this new column orthogonal to all prior output columns. Within each pass of the inner loop, we first cumulate in sum the length of the projection of the new column on the inner column. Recall that all prior columns are normalized to unit length, so we do not need to normalize the projection length. The second step in the inner loop is to subtract from the column icol its projection on the inner column, thus making columns inner and icol orthogonal. Last of all, we normalize this new column icol to unit length.

```
for (icol=1 ; icol<ncols ; icol++) {

  for (irow=0 ; irow<nrows ; irow++)      // Copy column icol from input to output
    output[irow*ncols+icol] = input[irow*ncols+icol] ;

  for (inner=0 ; inner<icol ; inner++) {   // Make icol orthogonal to all prior out columns
    sum = 0.0 ;
    for (irow=0 ; irow<nrows ; irow++)  // Compute dot prod of outputs icol and inner
      sum += output[irow*ncols+icol] * output[irow*ncols+inner] ;

    for (irow=0 ; irow<nrows ; irow++)  // Subtract from icol its projection onto inner
      output[irow*ncols+icol] -= sum * output[irow*ncols+inner] ;
    }

  // Make this output column unit length
  sum = 0.0 ;
  for (irow=0 ; irow<nrows ; irow++) {   // Sum the squared length of column icol
    temp = output[irow*ncols+icol] ;
    sum += dtemp * dtemp ;
    }
  sum = sqrt ( sum ) ;                      // This is the length now

  if (sum == 0.0)
    return 1 ;

  for (irow=0 ; irow<nrows ; irow++)      // Normalize to unit length
    output[irow*ncols+icol] /= sum ;
  }

return 0 ;
}
```

Putting It All Together

We have seen all of the core routines for performing forward selection with optional backward refinement. Now we discuss how to put them all together to produce a single routine that accomplishes this task. This will be illustrated with numerous code fragments, although the complete routine will not be shown here, as it includes too many distracting issues like the user interface and error handling.

The first issue is standardization of the universe of variables. It is critical to the algorithm that all variables be centered to have zero mean. Scaling to unit variance is not necessary, and in fact the authors of the paper cited at the beginning of this chapter do not do this. However, it is my own opinion that in most applications the variance of a variable is nothing more than an artifact of how it was measured and therefore just a source of confounding information. So I choose to rescale all variables to place them all on equal footing. If you would rather let them keep their original scaling so as to give them proportional weight in the selection process, feel free to do so. But even then, you *must* ensure that no variable is constant for all cases. A constant variable will cause numerical nightmares, not to mention that it makes no sense to even be present.

To clarify memory needs for the numerous variables that appeared in prior code listing, here are my allocations.

```
x = (double *) MALLOC ( n_cases * npred * sizeof(double) ) ;
kept_x = (double *) MALLOC ( n_cases * ncomp * sizeof(double) ) ;
cumulative = (double *) MALLOC ( npred * sizeof(double) ) ;
covar = (double *) MALLOC ( npred * npred * sizeof(double) ) ;
eigen_evals = (double *) MALLOC ( npred * sizeof(double) ) ;
eigen_structure = (double *) MALLOC ( npred * npred * sizeof(double) ) ;
work1 = (double *) MALLOC ( n_vars * sizeof(double) ) ;
work2 = (double *) MALLOC ( max_threads * ncomp * ncomp * sizeof(double) ) ;
work3 = (double *) MALLOC ( max_threads * ncomp * ncomp * sizeof(double) ) ;
work4 = (double *) MALLOC ( (max_threads * ncomp * ncomp + 2 * ncomp) *
                            sizeof(double) ) ;
work5 = (double *) MALLOC ( max_threads * ncomp * sizeof(double) ) ;
work6 = (int *) MALLOC ( max_threads * ncomp * sizeof(int) ) ;
kept_columns = (int *) MALLOC ( max_threads * ncomp * sizeof(int) ) ;
```

In those allocations, MALLOC() is my own error-checking memory allocation routine found in the file MEM64.CPP; you can substitute malloc() if you like living on the edge. We have n_cases rows in the x matrix and npred columns. (Array work1 is larger—n_vars— for a reason we'll see later.) The user has specified that we are to keep at most ncomp components and do backward refinement with at most max_threads threads.

We'll skip the standardization code and remind the reader that computation of the covariance (correlation due to standardization) matrix was shown on Page 10. We jump right to computation of the eigenvalues and vectors of the correlation matrix. This is not needed for our program, but it's nice because it allows us to provide the user with three informative results. First, we count the number of positive eigenvalues and use this to limit the number of components we compute. If there are fewer positive eigenvalues than the number of components the user requested, we must reduce that request due to the universe of variables matrix not having rank large enough to support that many orthogonal components. Second, if we compute a cumulative sum of the eigenvalues, we can print for the user the ordered variance contribution from the eigenvectors. (If this is not clear, please consult any source on principal components analysis.) Finally, we can print the eigenvectors of the data matrix. This may be a lot of information and may be of only limited interest, but in some applications, having a look at the eigenvectors can be quite revealing. This code is:

```
evec_rs ( covar , npred , 1 , eigen_structure , eigen_evals , work1 ) ; // EVEC_RS.CPP

sum = 0.0 ;
n_unique = 0 ;                    // Counts independent variates
for (i=0 ; i<npred ; i++) {       // We display cumulative eigenvalues
  if (eigen_evals[i] < 0.0)       // Rare; Happens only from tiny fpt errors
    eigen_evals[i] = 0.0 ;
  if (eigen_evals[i] > 1.e-9)     // 1.e-9 is arbitrary but reasonable
    ++n_unique ;
  sum += eigen_evals[i] ;
  cumulative[i] = sum ;
  }

for (i=0 ; i<npred ; i++)   // Make it percent
  cumulative[i] = 100.0 ∗ cumulative[i] / sum ;
```

We saw on Page 9 that the criterion for selecting the first variable is the mean squared correlation of the candidate variable with all other variables and that many users would be interested in seeing a table of these values. Here is how we compute them. Note that when we computed covar, we did only the lower triangle, because that's all that evec_rs() needs. But our criterion routine requires the entire matrix, so we copy the lower triangle into the upper triangle.

```
for (j=1 ; j<npred ; j++) {   // Copy to the upper triangle (covar is symmetric)
  for (k=0 ; k<j ; k++)
    covar[k*npred+j] = covar[j*npred+k] ;
  }

for (i=0 ; i<npred ; i++) {    // For each variable
  crit = newvar_crit ( npred , covar , 0 , kept_columns , i ,
                       work2 , work3 , work4 , work5 , work6 ) ;
  crit = (crit - 1.0) / (npred - 1) ;    // Crit included correlation with itself
  if (i == 0 || crit > best_sum) {    // Keep track of highest variable for stepwise if done
    best_sum = crit ;
    best_column = i ;                  // This will be the first variable selected
    }
  // Print the criterion (mean squared correlation) for user if desired
  }
```

Throughout the process, we will be inverting the **Z'Z** matrix, so we must make sure that the covariance matrix is nonsingular, which will guarantee that all submatrices are nonsingular. We reduce correlations enough to ensure that the smallest eigenvalue is positive. A single pass through this loop should virtually always be sufficient.

```
while (eigen_evals[npred-1] <= 0.0) { // Should require just one pass
  for (j=1 ; j<npred ; j++) {
    for (k=0 ; k<j ; k++) {
      covar[j*npred+k] *= 0.99999 ;
      covar[k*npred+j] = covar[j*npred+k] ;
      }
    }
  evec_rs ( covar , npred , 1 , eigen_structure , eigen_evals , work1 ) ;
  }
```

We now commence the loop that builds the subset of selected variables using forward selection and optional backward refinement. We count the kept variables in nkept. In the preceding code, we found the best variable, which is the first variable that we select. More explanations follow.

```
nkept = 1 ;   // Counts kept variables; will reach ncomp except in pathological cases
kept_columns[0] = best_column ;
best_crit = -1.e50 ;

while (nkept < ncomp) {        // While we still need to keep more components
   best_column = -1 ;          // Will flag if we got a good new column
   for (icol=0 ; icol<npred ; icol++) {   // Try all columns (predictors)

      // If this column has already been selected, skip it
      for (i=0 ; i<nkept ; i++) {
         if (kept_columns[i] == icol)   // Already in kept set?
            break ;
         }
      if (i < nkept)                     // True if already in kept set
         continue ;

      crit = newvar_crit ( npred , covar , nkept , kept_columns , icol ,
                     work2 , work3 , work4 , work5 , work6 ) ;
      if (crit > best_crit) {      // Did this variable just set a new record?
         best_crit = crit ;        // Update the best so far
         best_column = icol ;     // Keep track of which variable is best so far
         }
      } // For icol, finding the best variable to add to kept set

   kept_columns[nkept] = best_column ;    // Add this variable to the kept set
   ++nkept ;                               // Count variables kept

   if (type == 3) { // Backward refinement if requested
      if (max_threads > 1)
         while (SPBR_threaded ( npred , preds, covar , nkept , kept_columns ,
                           work2 , work3 , work4 , work5 , work6 , &crit )) ;
```

```
     else
        while (SPBR ( npred , preds, covar , nkept , kept_columns ,
                      work2 , work3 , work4 , work5 , work6 , &crit )) ;
     }

  } // While nkept < ncomp
```

The code just shown is straightforward, except for the optional backward refinement at the end of the loop. If the user has requested that multithreading be used, we call the multithreaded version of the refinement routine. Otherwise, we call the single-thread version.

What's important here is that the backward refinement is called inside a while() loop. If you look at the code for SPBR() that begins on Page 16 or the threaded version in the following section, you'll see that these routines return one if any variable in the kept set was replaced and zero otherwise. If a variable was replaced, another variable that previously held its own might now be vulnerable to replacement. Thus, we need to repeat the replacement process until we make a complete pass with no replacements.

This is a good time to elaborate on a possibly cryptic assertion regarding backward refinement made on Page 16. There, we asserted that when testing the last variable for possible replacement, if no replacement had been done to that point, we could safely skip testing the last variable. This should be clear for the first time SPBR() is called after adding a new variable: the last variable was just deemed to be the best, so it obviously is not subject to replacement.

But it may not be so clear that this still holds for subsequent calls of SPBR() inside the while() loop. Consider the state of things upon completion of the first call to SPBR(). Either replacement prior to the last column occurred, in which case the last column was also tested and replaced if appropriate, or no prior replacement occurred and the subset is still the same as when SPBR() was called. In the former case, the last column is now the best variable, and in the latter case, SPBR() returns zero so no further calls are done. Thus, every time SPBR() completes, we know that the last column is the best, and this fact holds recursively.

At this point we have a subset of forward-selected variables, possibly with backward refinement along the way. Regardless of where we go from here, we will need the Z'Z matrix for these selected variables. It's just a straightforward set of dot products identical to what we saw in the code on Page 10, except that there we used all variables and we divided by the number of cases to get a covariance/correlation matrix. Here we use only the kept set and we do not divide. See FSCA.CPP if this is not clear.

Components from a Forward-Only Subset

The final step in computing the components is different according to whether we included backward refinement in the subset selection process. Mathematically there need not be any difference, but in practice we want to treat the cases differently.

The reason for this difference is that if we employ strictly forward selection, we maintain ordering of the selected variables: the first variable captures the most variance in the universe, and so forth, in decreasing order of variance capture. If we have this ordering, we would like to preserve it in our computed components. This was discussed on Page 26. So we copy the selected variables from the universe and perform Gram–Schmidt orthogonalization in place. That subroutine leaves the columns normalized to unit length, but we want unit standard deviation (just a preference, but a good one). So we multiply everything by the square root of the number of cases to rescale the components.

```
for (i=0 ; i<n_cases ; i++) {
  for (j=0 ; j<nkept ; j++)
    kept_x[i*nkept+j] = x[i*npred+kept_columns[j]] ;
}

GramSchmidt ( n_cases , nkept , kept_x , kept_x ) ;

dtemp = sqrt ( (double) n_cases ) ;
for (i=0 ; i<n_cases ; i++) {
  for (j=0 ; j<nkept ; j++)
    kept_x[i*nkept+j] *= dtemp ;
}
```

We now have the standardized, orthogonal, ordered components in kept_x. Life is good. That may be enough for many users. But we lack one thing that would be good to have: the coefficients for converting the standardized original variables to the components. Many users will be curious to see them, and many may even want to use them to compute compatible components from a new dataset of the same variables, for example, to create an out-of-sample test set after using a training set to find the component weights. So let's tackle that.

We'll use ordinary linear regression to find the coefficients. Rather than using some sophisticated method like singular value decomposition, we just use the simple matrix inversion method. The **Z'Z** matrix is guaranteed nonsingular so we can invert it, and the time taken is insignificant compared to the compute time building the subset. We have

Z – n_cases by nkept matrix of selected standardized variables

M – n_cases by nkept matrix of orthogonalized components

We now wish to find the weight matrix **B** that will let us compute **M** from **Z** by the simple linear transformation shown in Equation (2.6).

$$\mathbf{M} = \mathbf{ZB} \tag{2.6}$$

The well-known least-squares solution to this problem is given by Equation (2.7).

$$\mathbf{B} = \left(\mathbf{Z}^{T}\mathbf{Z}\right)^{-1}\mathbf{Z}^{T}\mathbf{M} \tag{2.7}$$

Because the Gram–Schmidt process is a linear transformation, these predicted values for the original data will also be orthogonal except for minor floating-point errors. The code for doing this is as follows:

```
invert ( nkept , covar , work2 , &dtemp , work4 , work6 ) ;   // Inverse goes into work2

for (i=0 ; i<nkept ; i++) {          // Compute Z'M in work3.
  for (j=0 ; j<nkept ; j++) {
    sum = 0.0 ;
    for (k=0 ; k<n_cases ; k++)
      sum += x[k*npred+kept_columns[i]] * kept_x[k*nkept+j] ; // x is Z, kept_x is M
    work3[i*nkept+j] = sum ;
  }
}

for (i=0 ; i<nkept ; i++) {          // Multiply those two matrices to get B, the coefficients.
  for (j=0 ; j<nkept ; j++) {        // They will go into work4
    sum = 0.0 ;
    for (k=0 ; k<nkept ; k++)
      sum += work2[i*nkept+k] * work3[k*nkept+j] ; // work2 is inverse, work3 is Z'M
    work4[i*nkept+j] = sum ;
  }
}
```

Components from a Backward Refined Subset

If we used backward refinement in the subset generation process, we lost ordering of the variables. The first (originally best) variable selected may not have even survived refinement! In practical applications, that changes how we may want to handle component generation.

The authors of the paper cited at the beginning of this chapter take a reasonable approach. They treat the selected subset of variables as a universe and perform forward-only selection from them, followed by the same Gram–Schmidt orthogonalization that we presented in the prior section. This results in orthogonal components that are ordered according to variance capture. However, *this ordering is relative to the selected subset only, not the original universe of variables.* Of course, the selected subset is itself optimal in a perfectly reasonable sense, so deriving variance-capture ordering from them is also reasonable.

Still, it is my own opinion (and this is strictly a matter of opinion) that this can be misleading to the user. After all, it can and often does happen that the best single variable derived from the universe is entirely different from the best single variable derived from a subset selected with backward refinement. And naturally this deviation occurs right on down the line. In extreme cases, it could happen that *many* of the best forward-selected variables, selected in regard to the entire universe, are not represented in the variables that pop out of forward selection from a backward refined subset. Reporting ordering of backward-refined variables, even if derived from secondary forward selection, grates on my intuition to some degree. It would be too easy to mislead the user.

For this reason, I take a radically different approach to computing ordered orthogonal components from a set whose selection included backward refinement. I compute the principal components of the selected subset. This gives us ordering, with the first component capturing the most variance from the selected subset, the second component capturing the second-most variance, and so forth. But the ordering has nothing to do with the variables. Rather, we thereby perform what our intuition at the start of this chapter directed: we have computed principal components for the dataset but using only a small subset of the universe.

Here is the code for doing this. Recall that earlier we computed $\mathbf{Z'Z}$ in covar. But we need a covariance/correlation matrix, so we must divide by the number of cases. The eigenstructure routine requires only the lower triangle of this symmetric matrix. Then we compute the eigenstructure and, for the user's edification, compute the cumulative percent of the total unique variance of the universe that we capture with successive principal components. We computed n_unique early in the process, as shown on Page 30.

```
for (j=0 ; j<nkept ; j++) {
  for (k=0 ; k<=j ; k++)
    covar[j*nkept+k] /= n_cases ;
  }

evec_rs ( covar , nkept , 1 , eigen_structure , eigen_evals , work1 ) ;

sum = 0.0 ;
for (i=0 ; i<nkept ; i++) {      // We display cumulative eigenvalues
  if (eigen_evals[i] < 0.0)   // Rare, if ever; Happens only from tiny fpt errors
    eigen_evals[i] = 0.0 ;
  sum += eigen_evals[i] ;
  cumulative[i] = sum ;
  }

for (i=0 ; i<nkept ; i++)      // Make it percent
  cumulative[i] = 100.0 * cumulative[i] / n_unique ;
```

We can do two different things with the eigenvectors, each of which (a column in the matrix) is normalized to unit length by evec_rs(). If we divide each by the square root of its corresponding eigenvalue, we get the weights for computing the component standardized to unit variance. We printed weights in the prior section, and so are tempted to do the same here. However, if instead of dividing we multiply each eigenvector by the square root of its eigenvalue, we get the correlation between this component and the corresponding variable. This can be even more informative for the user, and if the user requires standardized components, these weights can be trivially computed by dividing the correlation by the eigenvalue. For this reason, I favor printing the correlations. Take your choice.

An Example with Contrived Variables

Here is an example of each of the two algorithms (strictly forward, and forward with backward refinement). For this example, the following nine variables are employed:

> **RAND1–RAND6** are independent (within themselves and with each other) random time series
>
> **SUM12 = RAND1 + RAND2**
>
> **SUM34 = RAND3 + RAND4**
>
> **SUM1234 = SUM12 + SUM34**

When we run the FSCA algorithm in *VarScreen* using the option for strict ordering (no refinement), we first see the following results printed:

There are 6 unique (non-redundant) sources of variation
The number of components computed is therefore being reduced
to this value.

Eigenvalues, cumulative percent, and principal component factor structure

Eigenvalue	2.988	1.986	1.052	1.015	0.987	0.972
Cumulative	33.195	55.263	66.957	78.240	89.203	100.000
RAND1	0.4835	0.4964	-0.6476	-0.1497	-0.1080	-0.2576
RAND2	0.4597	0.5206	0.6390	0.1478	0.1037	0.2770
RAND3	0.5246	-0.4808	-0.0470	-0.2077	0.6690	-0.0271
RAND4	0.5175	-0.4859	0.0620	0.2194	-0.6661	0.0240
RAND5	-0.0198	-0.0198	-0.4669	0.4999	0.1474	0.7139
RAND6	0.0020	0.0260	0.0233	0.7937	0.2265	-0.5635
SUM12	0.6800	0.7331	-0.0090	-0.0021	-0.0036	0.0128
SUM1234	0.9997	0.0239	0.0012	0.0040	0.0003	0.0073
SUM34	0.7331	-0.6800	0.0104	0.0076	0.0039	-0.0023

We have nine variables in the universe, but the program notes that there are only six unique sources of variation. This is not surprising, because the three sum variables are just combinations of the other variables. Since by definition the computed components must be independent, the program limits us to just six components.

The first component accounts for one-third of the total variation in the dataset, and it correlates almost perfectly with SUM1234, very highly with SUM12 and SUM34, moderately highly with RAND1-RAND4, and not at all with RAND5 or RAND6. None of this should be surprising.

The second component is just the contrast between RAND1 and RAND2 versus RAND3 and RAND4. In conjunction with the first component, it gives us over 55 percent of the total variation. The remaining components are other contrasts as well as RAND5 and RAND6.

Next, we get a list of the mean squared correlation of each variable in the universe with all other variables:

Mean squared correlation of each variable with all others

RAND1	0.091
RAND2	0.088
RAND3	0.096
RAND4	0.095
RAND5	0.000
RAND6	0.000
SUM12	0.181
SUM1234	0.248
SUM34	0.191

It is not surprising that RAND1–RAND4, along with their various sums, have positive mean squared correlations, while RAND5 and RAND6 have zero correlations.

Last of all, we get the table of coefficients needed to compute the six components from the chosen six variables in the subset. The variables appear in order of importance. Note that each component depends on only the corresponding ordered variable and all previously selected variables. This is a direct result of the Gram–Schmidt orthogonalization process described on Page 26. This table was computed by the algorithm on Page 34, ultimately ending up in work4.

Variable	1	2	3	4	5	6
SUM1234	1.0000	-0.9730	0.0181	0.0106	0.0047	-1.4045
SUM12	-0.0000	1.3953	-0.9696	-0.0091	-0.0131	0.9888
RAND2	0.0000	-0.0000	1.3842	-0.0129	0.0380	-0.0081
RAND6	0.0000	0.0000	-0.0000	1.0001	-0.0169	-0.0071
RAND5	0.0000	-0.0000	0.0000	0.0000	1.0007	-0.0017
RAND4	-0.0000	0.0000	-0.0000	-0.0000	-0.0000	1.4188

Observe that the best single variable for reproducing the entire universe of values is SUM1234, the sum of four other variables in the universe, and the first component is just this one variable (its coefficient is 1.0 and all other coefficients are 0.0).

The second variable selected is another sum variable, and the corresponding component's value is computed as that sum variable times 1.3953, minus the prior selected variable times 0.9730.

The third variable selected is a similar weighted sum, primarily based on RAND2. The next two components are essentially equal to the two completely independent variables, RAND6 and RAND5. Note that their coefficients are almost exactly 1, and all other coefficients are almost exactly 0. And the last component is a complex mix of other variables.

We then use this same universe of variables to demonstrate the other FSCA option, forward selection combined with backward refinement. The initially printed information (eigenstructure and mean squared correlations) are the same as in the prior example, so we will skip straight to the interesting part, the log of variables being added and replaced:

```
Commencing stepwise construction with SUM1234
Added SUM12 for criterion=4.973085
   Replaced SUM1234 with SUM34 to get criterion = 4.973123
Added RAND2 for criterion=6.011605
   Replaced SUM12 with RAND1 to get criterion = 6.011623
Added RAND6 for criterion=7.011701
Added RAND5 for criterion=8.010402
Added RAND4 for criterion=8.999919
   Replaced SUM34 with RAND3 to get criterion = 8.999940
```

As in the prior example, the first variable selected is SUM1234. We then add SUM12, as in the prior example. (Both options will always select the same first two variables.) But then something interesting happens: SUM1234 is replaced by SUM34, giving us a two-variable set of SUM12 and SUM34. To me, this is prettier than SUM1234 and SUM12.

We then add RAND2, which immediately triggers the replacement of SUM12 with RAND1. After that, we add the two totally independent variables, RAND6 and RAND5. Finally, we add RAND4, which triggers the replacement of SUM34 with RAND3. The final results are as follows:

Eigenvalues, cumulative percent, and selected principal component factor structure

Eigenvalue	1.056	1.023	1.002	0.988	0.983	0.948
Cumulative	17.600	34.646	51.343	67.811	84.194	100.000
RAND3	0.2269	0.5167	0.2772	0.5747	-0.5221	-0.0431
RAND1	0.5751	0.0508	-0.1047	0.3593	0.5829	0.4322
RAND2	-0.6877	0.0094	0.1597	0.2264	0.0044	0.6709
RAND6	-0.0862	-0.6254	0.4725	0.4844	0.1757	-0.3357
RAND5	0.4228	-0.5366	0.1187	-0.2014	-0.5355	0.4381
RAND4	0.1210	0.2723	0.8070	-0.4496	0.2300	0.0702

The final set of selected variables is intuitively more appealing than what we got with the strict ordering option, because it's just the individual random variables, without their various sums. Because replacement has destroyed any ordering of the subset, it makes the most sense to me to just compute the components as the principal components of the final subset. Note that the eigenvalues are all nearly equal, meaning that the components have no strong ordering either, which is exactly what we would expect when the variables are themselves independent. Also note that the values in the table are the correlations between the components and the variables, and they can be converted to weights by dividing each column by the eigenvalue at the top of the column.

CHAPTER 3

Local Feature Selection

Intuitive Overview of the Algorithm

Most common feature selection algorithms are primarily oriented toward favoring features that are at least somewhat predictive over the *entire* domain of the feature set. This predictivity may be nonlinear, and it may interact with other features, but such a predictor will be at a significant advantage over more powerful but only locally predictive candidates if the nature of its relationship to a target variable is at least somewhat consistent across the domain of all possible values of all candidate features.

This global favoritism can be a major problem, because modern nonlinear models can obtain a lot of useful predictive information from variables whose power is limited to small areas of the domain or whose predictive relationship changes significantly over the domain. But if our predictor selection algorithm fails to find such variables, focusing instead on more global candidates, we lose out on what may be valuable information.

For example, consider the simple XOR problem. Suppose we have two standard normal random variables and we define two classes. A case is a member of Class 1 if both of our variables are positive or both negative, and it is in Class 2 if one is positive and the other negative. This classification problem can be solved with 100 percent accuracy by a simple rule, and modern nonlinear models should have no trouble achieving nearly perfect performance. Yet if we were to augment these two variables with numerous other worthless predictor candidates and then try to identify the two true predictors, an amazing number of otherwise sophisticated predictor selection algorithms would fail to find them. Not only are the marginal distributions of both variables identical in both classes, but the relationship of each variable to the class depends completely on the value of the other variable, with the relationship reversing across the domain. This is a tough problem.

© Timothy Masters 2020
T. Masters, *Modern Data Mining Algorithms in C++ and CUDA C*,
https://doi.org/10.1007/978-1-4842-5988-7_3

This same issue arises in applications that are closer to reality. For example, a common phenomenon in equity market prediction is that certain families of indicators have considerable predictive power in times of low market volatility but become useless in times of high volatility. The presence of a large amount of high-volatility data in the dataset dilutes the predictive power of such variables and may put otherwise excellent indicators at a competitive disadvantage. And this problem arises in many other applications. The effectiveness of medical treatments can vary according to the patient's age, weight, and a potentially large number of other unknown conditions. Identification of vehicles and pedestrians by a self-driving car's control system can depend on features that are vital in some contexts and distracting clutter in others. We need a feature selection algorithm that is sensitive to predictive power that comes, goes, and even reverses, according to location in the feature domain.

In terms of modeling, we can deal with inconsistent behavior in a large predictor set by using sophisticated nonlinear models (which are prone to overfitting!) or by using different models in differing regimes (assuming that we know how to define these regimes!). But consider the premodeling stage, when we are searching for predictor candidates. We would like to have a predictor selection algorithm that can automatically find such regime-dependent behavior and identify powerful predictors, even if this power is localized.

The feature selection algorithm described in "Local Feature Selection for Data Classification" by Narges Armanfard, James P. Reilly, and Majid Komeili (*IEEE Transactions on Pattern Analysis and Machine Intelligence*, June 2016) fits the bill nicely. We'll now present a condensed and intuitive overview of how it operates.

There are a large number of possible approaches to feature selection. You've doubtless seen some based on mutual information and uncertainty reduction, techniques that are effective at detecting highly nonlinear relationships. Some other techniques actually train predictive models and perform their feature selection by intelligently choosing inputs for these models. Early discriminant analysis methods involve the use of Mahalanobis distances to find dimensions of maximum separation when the predictors are highly correlated, optimally taking correlation into account. The LFS algorithm presented here is based on yet another approach, a concept akin to nearest neighbor classification, but taken to a much higher level of sophistication.

We begin with a simple example: we want to predict success in college, with students divided into two classes: those who graduate and those who drop out. We measure four candidate predictors for each student in our study dataset and standardize the values of these predictors (mean zero and standard deviation one) to put their variation on a level playing field. These candidate predictors are

1) SAT score

2) High school grade point average (GPA)

3) Circumference of thumb divided by circumference of index finger

4) Day of month student was born

Suppose we randomly choose two students, both in the *Success* class. For each of these four features, think about the average difference in predictor values we would see for these two successful students. Because the students are in the same class, their expected difference for each of the four predictor candidates would be relatively small. But now suppose we randomly choose two students, one in the *Success* class and one in the *Dropout* class. The expected difference between these two students would be about the same as it was for the "same class" students for the third and fourth candidate predictors but much larger for the first and second candidate predictors. This is because a person who graduated would probably have a higher GPA and SAT score than a dropout, leading to a large difference, while these two students would probably have similar finger sizes and birthdays, at least relatively speaking.

If we effectively estimated these expected differences throughout the dataset, looking at every pair of students, we would conclude that the first two candidate predictors are the ones we want, because the expected difference in these two features for students in different classes greatly exceeds the expected difference for students in the same class, while for the third and fourth candidate predictors, we observe about the same difference, regardless of whether the two students are in the same class or different classes.

Now, instead of looking at candidate predictors individually, let's look at them in pairs: 1 and 2, 1 and 3, and so on. A good measure of the difference between two cases is the Euclidean distance between them. Let $x_m^{(i)}$ represent the value of variable m as measured for case i, and let the vector $x^{(i)}$ represent the set of all variables for this case. Then the distance between case i and case j is given by Equation (3.1).

$$d_{ij} = \left\| x^{(i)} - x^{(j)} \right\| = \sqrt{\sum_m \left(x_m^{(i)} - x_m^{(j)} \right)^2} \qquad (3.1)$$

It should be clear that the pair of variables consisting of the first two competitors will have the greatest expected inter-class distance between cases, the pair consisting of the last two competitors will have the least expected inter-class distance, and mixed pairs will have intermediate values.

Intuition can now guide us toward a good way to choose an effective set of candidate predictors. We look for a set that has a high contrast between expected intra-class distance (which we want to be small) and inter-class distance (which we want to be large). Neither quality alone is good. For example, if we find a set of candidates that produces large average inter-class separation between cases but the expected separation between cases in the same class is also large, we have gained nothing; we cannot look at either quality in isolation. We must find a balance, a way to trade off the desirable quality of low intra-class separation with the also desirable quality of high inter-class separation. The LFS algorithm has an automated way to find the optimal tradeoff, a topic which we will return to later.

All that we've seen so far is good, and the algorithm just outlined would work fairly well in practice. However, it is missing the "Local" component of the "Local Feature Selection" algorithm. We still need a way to handle the problem of predictive power that varies across the feature domain. For example, the distribution shown in Figure 3-1 would foil the algorithm just described.

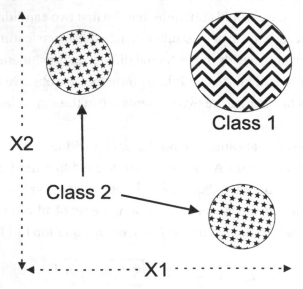

Figure 3-1. *A job for Local Feature Selection*

In this example, we have two classes, one of which is split into two distinct subsets. Think about how the variable selection algorithm just described would perform when presented with this problem. Half of the cases in Class 2 would have excellent inter-class separation from Class 1 via X1, though no separation at all via X2. The other half would experience the opposite behavior, gaining great separation via X2 but none via X1. If inter-class separation were the only consideration, the algorithm would pick up X1 and X2 easily, even though different variables are responsible for the inter-class separation; having just one of the summands in Equation (3.1) large is good enough to make the distance large.

The killer lies with the intra-class separation. Pairs of cases that both lie within the same subset of Class 2 would have nicely small separation. But if one case in Class 2 lies in one subset and the other case lies in the other subset, the distance between them would be enormous, larger even than the inter-class separation! So the average intra-class separation for Class 2 would be so large that it would be nearly commensurate with the inter-class separation. It's unlikely that (X1,X2) would stand out among competitors as a set of effective predictors, even though this figure shows that they are fabulous.

The key element of the paper cited at the beginning of this chapter is that the problem shown in Figure 3-1 can be alleviated by weighting the distances with intelligently computed weights. The primary focus in the weighting scheme is that pairs of cases which are close are given higher weights than pairs which are distant, with the weighting dropping off exponentially with distance. It's somewhat more complicated than that, because the class memberships of the cases are taken into account, as well as the global distribution of the distance metrics. We'll delve into those details later in the chapter. For now, just understand that the algorithm for computing sensible weights is quite clever and effective.

In order to get an idea of what's happening in regard to weights, the four histograms in Figure 3-2 show the weights generated from a test with data having the pattern shown in Figure 3-1.

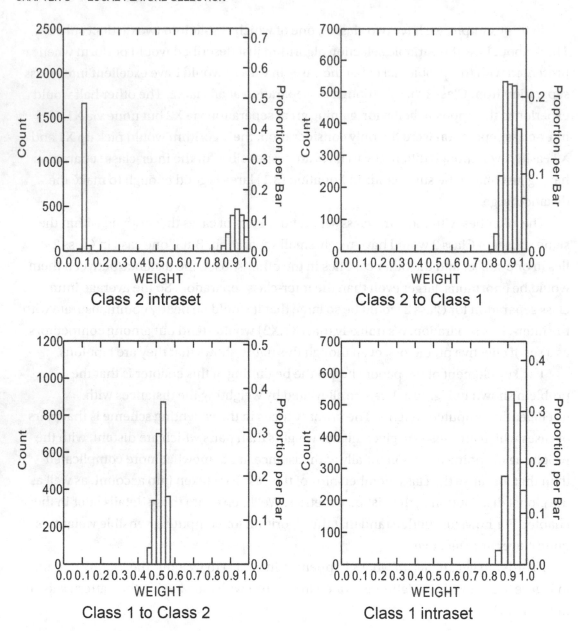

Figure 3-2. *Weights for split-class example*

The most interesting of these four histograms is the upper-left, which shows the weights for pairs of cases that are both in Class 2. We see that half of the weights are clustered near the maximum possible weight, 1. These are the pairs of cases that are both in the same subset of Class 2. The other half of the weights are clustered near zero, the minimum possible. These are the pairs of cases that, while both in Class 2, are in different subsets of this class. So we see that when the intra-class separation (mean distance separating cases that are both in Class 2) is computed with weighted distances, pairs that span the two subsets are downplayed, thus providing a more realistic estimate of the intra-class separation.

The Class 1 intra-class weights are all close to 1 because this class is not split into subsets. Also, when we are considering cases in Class 2 and looking at their distances from cases in Class 1, we have full weighting. The weights are about 0.5 when we consider cases in Class 1 and look at the distances to cases in Class 2 (the weighting algorithm is asymmetric). Very roughly speaking, this is because there are two possible ways the difference can go. Later in this chapter, when we study the weight equations, we will see exactly how this comes to be.

What This Algorithm Reports

Because the algorithm performs optimal-candidate selection separately for each case, there is no practical way to report a single optimal candidate set, let alone a sorted list of subsets like we are able to achieve with some other algorithms. Instead, it counts the number of times each candidate predictor appears in an optimal subset. For example, we might see that X2, X7, and X35 form an optimal subset at some point; X3, X7, and X21 form another optimal subset; X7 and X94 form another; and so forth. X7 appeared in an optimal subset three times, while each of the other subset members appeared just once. So it looks like X7 is on its way to becoming popular and heading up the importance list.

This does not mean that X7 alone is valuable. In fact, it may be (and often is!) that X7 alone is worthless; its value is only in conjunction with other candidates. This is why LFS is superior to many other feature selection algorithms, which often rely on some form of stepwise selection and hence may ignore individually worthless candidates. But this property of reliance on other predictors is not a problem. The reason is that most modern prediction models, if given a list of the most popular predictors, can sort out the complex relationships between them and perform well. All they need is preprocessing to weed out the worthless candidates, so they are not overwhelmed. This sort of preprocessing is an ideal application for Local Feature Selection.

A Brief Detour: The Simplex Algorithm

The LFS algorithm makes extensive use of a standard optimization procedure called the *Simplex Algorithm*, so it's reasonable to take a brief detour from the primary topic and present an overview of this important algorithm and its implementation here. This is especially true because some readers may wish to use SIMPLEX.CPP for their own applications apart from its use in the *VarScreen* program. This implementation is not as broadly applicable as highly sophisticated versions in that it lacks three advanced features. However, these features are not required by the LFS algorithm, and in my opinion, they are minor omissions for most common applications. For readers who are already familiar with the simplex algorithm, know that these limitations are as follows:

1) Strict equality constraints are not supported, only less-than and greater-than constraints.

2) All basis vectors are explicitly stored, rather than being simply flagged. This results in greater memory requirements, although in this era of vast storage capacity this is of little or no consequence. Moreover, flagging instead of storing basis vectors vastly increases the complexity of the algorithm, so this tradeoff was a no-brainer.

3) No provision is made for degenerate conditions that can result in endless cycling, although this condition will be reported if it occurs. However, in practical problems that are well formulated, cycling is rare. And because the program is highly commented and cleanly structured, interested readers should have no problem adding cycle-breaking code if desired.

The mathematical foundations of the simplex method for linear programming are widely available and hence will not be presented in detail here. This section will discuss the fairly general form of the linear programming problem that the Simplex class solves, and we will show how the class can be incorporated into other programs.

The Linear Programming Problem

The most basic linear programming problem involves maximizing a linear function (the *objective function*) subject to constraints of the "less-than-or-equal" variety. An example objective function is that shown in Equation (3.2), and two example constraints are shown in Equations (3.3) and (3.4).

$$z = c_1 x_1 + c_2 x_2 + c_3 x_3 \tag{3.2}$$

$$a_{11} x_1 + a_{12} x_2 + a_{13} x_3 \leq b_1 \tag{3.3}$$

$$a_{21} x_1 + a_{22} x_2 + a_{23} x_3 \leq b_2 \tag{3.4}$$

In addition to the user-specified constraints, linear programming always imposes two other families of constraints:

1) All of the x variables must be nonnegative.

2) All of the b limits must be nonnegative.

There are no restrictions placed on the a or c coefficients.

Note that the number of problem constraints (2 in this example) may be less than, equal to, or greater than the number of variables (3 here). The nature of this relationship has no impact on the method for solving the problem.

The form of the problem just shown can handle many real-life linear programming problems. The restriction that each x be nonnegative rarely has an impact, because the x values often represent physical quantities of materials. The restriction that each b be nonnegative and the inequality be "less-than-or-equal" also rarely has an impact, because the a coefficients often represent cost of those materials, with each b representing a limit on total cost. However, there are two situations that this form can almost, but not quite, handle:

1) We may want to have one or more b values be negative.

2) We may want one or more of the inequalities to be "greater-than-or-equal."

These two situations are really just two sides of the same coin, because multiplying both sides of an inequality by –1 flips the sign of b and reverses the inequality. So if we relax the "<=" inequality restriction to also allow ">=", we can handle both of those situations. Moreover, the restriction that the x values be nonnegative then goes away in practice because we can always flip the sign of every coefficient involving an x that is nonpositive and thereby create a new x that is nonnegative.

There is one more generalization that is useful in some problems: a strict equality constraint. This ability is not needed for LFS, so I did not bother implementing it. However, readers who brush up on the algorithm should be able to modify my highly commented code to handle this situation.

Interfacing to the Simplex Class

You will need to create two arrays. One holds the objective function coefficients. By convention, we say there are N of them, being the coefficients of x_1, x_2, ..., x_N. The other array is really a matrix strung out with columns changing fastest. There is one row for each of the M constraints, and each row has $N+1$ columns. In each row, the first element is the b limit for that constraint, and the remaining N columns are the x coefficients for that constraint. For the example just shown, this constraint matrix/array would be b_1, a_{11}, a_{12}, a_{13}, b_2, a_{21}, a_{22}, a_{23}. There are a total of $M(N+1)$ elements. If the constraints contain both "<=" and ">=" inequalities, all of the "<=" constraints must come first. We say that there are M_{LE} of them.

The first step is to create a new Simplex object. The constructor takes four integer parameters:

> N – The number of variables.

> M – The number of constraints.

> MLE – The number of constraints of the "<=" type; it may be 0 through M.

> $Print$ – If nonzero, detailed computational steps are printed to MEM.LOG.

```
sptr = new Simplex ( N , M , MLE , Print ) ;
```

We use separate calls to set the x objective coefficients vector and the constraint matrix. The order in which they are called does not matter.

```
sptr->set_objective ( coefs ) ;
sptr->set_constraints ( constraint_matrix ) ;
```

We can then compute the optimal values and retrieve the results. The parameters are shown below those lines.

```
sptr->solve ( max_iters , eps ) ;
sptr->get_optimal_values ( &optval , x ) ;
```

max_iters – This is a safety valve. In most practical applications, convergence is obtained within $\max(M,N)$ iterations, although in unusual situations it can go higher. In fairly rare pathological situations, the algorithm can even get locked in an endless loop. There are complex methods for avoiding this, but the guaranteed loop-free algorithms are cumbersome, as well as slow and inefficient. Given that endless looping is practically unknown in well-formulated real-life applications, I keep to the traditional method in this code. So if you set max_iters to several times max(M,N), you will almost certainly be safe. In the unlikely event that your problem is caught in an endless cycle, the solve() routine will fail with an error code, as will be discussed soon.

eps – There are several places in the code where a value needs to be tested to see if it equals zero. Accumulated floating-point errors can result in values that theoretically should be zero to instead be tiny nonzero numbers. When a value that should be zero is not, bad things can happen. This parameter lets us relax our definition of zero and call a tiny number zero. I typically set it to about 1.e-8. This may rarely cause iteration to stop slightly before the true maximum is reached, or even more rarely cause the system to be falsely labeled as unbounded. (Unbounded means that the objective function can increase to infinity without violating the constraints.) These problems arising from a relaxed zero are rare and generally innocuous, while the problems from not properly accommodating inevitable floating-point errors can be devastating.

optval – This is the computed maximum value of the objective function.

x – The user supplies a pointer to a vector N long. The values of the variables that produce the maximum objective are copied to this vector.

The solve() routine returns an integer with the following meaning:

0 – Normal return, maximum found.

1 – Objective function is unbounded.

2 – Too many iterations without convergence (may indicate infinite-loop cycling).

3 – Conflicting constraints prevent a feasible solution.

4 – Constraint matrix is not full rank (there is redundancy).

Note that it is legal to reuse a Simplex object. After solving a problem, you may call set_objective() and/or set_constraints() again, followed by solve() for the new problem.

If you are interested in checking the veracity of the results, two routines are available. Both return 1 if there is an error, 0 otherwise. If there is an error, the error parameter returns the numerical value by which the result is in error. The eps parameter should be a small number, perhaps 1.e-8, which allows for a small fudge factor in flagging an error. No algorithm involving numerous floating-point calculations can be expected to precisely match theoretical expected results.

int Simplex::check_objective(double *coefs , double eps , double *error);
coefs is the original coefficient matrix defining the objective function.

The check_objective() routine is a good way to assess the degree to which cumulated floating-point errors have impacted results. This routine forms the dot product of the objective function coefficients with the computed optimal values of the x variables. It compares this number with the final optimal value of the objective function, which is computed and updated every step of the way. In other words, these two numbers, which should theoretically be identical, are computed in completely different manners, with the value returned by get_optimal_values() being the end product of a large number of floating-point operations as the optimal value was found.

int Simplex::check_constraint (int which , double *constraints , double eps,
double *error) ;

which ranges from 0 through $M-1$ and specifies which of the M constraints is checked.

constraints is the original constraints matrix, already described.

A Little More Detail

If all you want to do is interface the Simplex class with your own program, you've already seen all that you need to know; you can safely skip this section. On the other hand, if you're curious about the Simplex algorithm and would like just a little more information about it, read on to get an explanation that is only slightly dumbed down to a level suitable for mathematically challenged readers. On the third hand, if you want deep details about the simplex method, you're in the wrong place. There is much material available online, and some of it is even pretty good. Best of all, buy a good textbook on the subject. *Linear Programming* by Robert Vanderbei is not for the faint of heart, but for the serious student I don't think there's a better reference.

A key to understanding the simplex algorithm is to know that the (properly defined) constraints form a convex structure consisting of vertices connected by lines. The algorithm begins at a vertex that satisfies all of the constraints (called a *feasible solution*) and then jumps from vertex to vertex, (usually) steadily increasing the objective function with each jump. Eventually a point is reached at which the objective function has reached its maximum value. This is the optimal solution.

There are three primary things that can go wrong with this rosy view of the situation:

1) The constraints might be in fundamental conflict such that it is impossible to satisfy all of them simultaneously.

2) The constraints might not form a closed structure that bounds the objective function, with the result that one or more of the x variables can increase without bound and similarly drive the objective function toward infinity.

3) The basic simplex algorithm may find itself in an endless loop in which the objective function never increases and the path from vertex to vertex endlessly repeats the same pattern.

The first two of these problems are the making of the user and can be corrected only by proper reformulation of the problem. The third problem is a defect of the simplex algorithm itself, and the Simplex class provided here does nothing to solve it. However, be comforted by the fact that this sort of disastrous cycling is quite rare in practical problems. The sad truth is that cycle-resistant simplex algorithms are not only a lot more complex than the algorithm given here, but generally they run much more slowly and hence are most suited for last-resort situations. In any case, you can discover if cycling is occurring in your application by setting the iteration limit max_iters to a very large number and seeing if solve() still fails with a "too many iterations" error.

I'll quickly mention one more interesting item about the simplex algorithm, just enough to wet the lips of an eager student. It was stated earlier that the simplex algorithm begins at a *feasible solution* and advances the objective function upward from there. A feasible solution is one that satisfies all of the constraints, though it may be far from optimal. If all of the constraints are of the <= variety, it's easy to find a feasible solution at which to begin iterating: just set all of the x values to zero. Since the b limits must all be nonnegative, the constraints are automatically satisfied.

But if any of the constraints are of the >= variety, we are in trouble, because it's rare that a feasible solution will be obvious. In this case, we must solve the problem in two steps (*phases*). In the first phase, we modify each >= constraint by introducing an additional variable with a coefficient of +1 and whose initial value is equal to the b limit for that constraint. In this way the constraint is satisfied. We temporarily change the objective function to be the negative sum of any such additional variables and maximize this sum. If the constraints are not in conflict, we will be able to drive this negative sum to zero, implying that any additional variables now have a value of zero and hence can be removed. This leaves us with the original problem that we want to solve, along with an initial feasible solution from which the second phase of simplex optimization can begin.

A More Rigorous Approach to LFS

Complete source code for the LFS algorithm can be found in the files LFS.CPP (the main coordinator of operations), LFS_DO_CASE.CPP (coordinates processing a single case in a single thread), LFS_WEIGHTS.CPP (computes the case weights), and LFS_BETA.CPP (computes the optimal beta weight for balancing the importance of intra-class and inter-class separation). For the remainder of this chapter, we will present the LFS algorithm in moderate mathematical detail (though with not quite such rigor as in the paper cited at the beginning of this chapter). The strictly correct notation used in the paper can be confusing and cumbersome at times, so I have taken some (hopefully forgivable) liberties at times when I believe that relaxing rigor makes the intuition behind operations more clear. Code snippets illustrating the various operations will be interspersed, along with occasional complete subroutines.

Here are a few summary points about the algorithm that must be firmly understood:

- The aspect of the LFS algorithm that can be most difficult to grasp at first, and which I will therefore emphasize now, is this: every case in the dataset is processed separately and independently. All of the operations that form the core of the algorithm can be done in parallel (and will be in my implementation), with no interaction among them. In other words, we solve the bulk of the problem for the first case. While we're doing that, we can (if we wish) also be solving the bulk of the problem for the second case, with no need to look at intermediate or final results for the first case. Et cetera. So for the vast majority of the ensuing discussion, we will have a single fixed case i in mind, which I call the *current case*, and all of our operations will revolve around how this one case, whose data values are in the M-element vector $x^{(i)}$, interacts with the other $N–1$ cases in the dataset.

- Our goal when processing the current case i is to find a binary vector $\mathbf{f}^{(i)}$ whose M elements flag which variables are the best performers for this case. In other words, we want the best possible classifier for this one case.

- Roughly speaking (more detail in the next bullet point), we define a "good classifier" (a good binary feature vector $\mathbf{f}^{(i)}$) for this "current" case as one that has two properties:

  1. The average distance between this case and every other dataset case that is in this class is small.

  2. The average distance between this case and every other dataset case that is *not* in this class is large.

  Look back at the college success example on Page 45. For most, if not all cases in this dataset, we would find that the best binary feature vector is {1, 1, 0, 0}. This is because the first two predictor candidates would likely produce a relatively large distance (Equation (3.1)) between graduates and dropouts, while the other two predictor candidates would likely show about the same distance, regardless of whether the students being compared are in the same class or different classes.

- The only problem with the prior bullet point is that it does not take local behavior into account. Recall the issue we had with the example shown in Figure 3-1 on Page 46. Our definition of a good binary feature vector would be subverted by the enormous average distance between members of Class 2 due to its splitting into distinct subsets. Therefore, when we compute the average distances, we need to weight the terms in such a way that we favor cases in the local neighborhood of case i, the case currently under consideration.

- The prior two bullet points leave us with a bit of a chicken-and-egg dilemma: in order to compute weights, we need a definition for the neighborhood of case i, which in turn depends on the variables that go into computing distances (those variables that are flagged as "1" in $\mathbf{f}^{(i)}$). But we can't evaluate the quality of a trial $\mathbf{f}^{(i)}$ until we know the weights. Our only choice is to use an iterative procedure: begin with a "neutral" $\mathbf{f}^{(i)}$ and use it to estimate weights. Then use those weights to get a better $\mathbf{f}^{(i)}$. Then use that better $\mathbf{f}^{(i)}$ to compute new weights. Repeat as needed. Fortunately, much empirical evidence indicates that convergence to high-quality weights and inclusion flags is fast, becoming decent after just two iterations and very good after a third.

We can now put all of these points together into a roughly stated general algorithm. It looks like this:

1) Initialize $\mathbf{f}^{(i)}$ to all zeros (no variables selected) for all i. This is an N cases by M variables matrix of zeros.

2) For a small number of iterations (typically 2 or 3).

3) $\mathbf{f}_{\text{prev}} = \mathbf{f}$ (copy the current N by M binary inclusion matrix to a "prior value" work area).

4)   For i from 1 to N (process all cases individually).

5)   Using $\mathbf{f}_{\text{prev}}$, compute the vector of N weights for this case i.

6)   Compute ε, the maximum possible inter-class separation.

7)   For each of about 10–30 trial values of β in $(0, 1)$.

8)   Compute a real-valued M-vector $\mathbf{f}_{\beta,\text{real}}^{(i)}$ that minimizes a proxy for the intra-class separation, subject to the constraint that the proxy inter-class separation is at least $\beta\varepsilon$.

9)   Use a random process to find a binary approximation $\mathbf{f}_{\beta}^{(i)}$ to $\mathbf{f}_{\beta,\text{real}}^{(i)}$ that, of the valid random tries, minimizes the actual intra-class separation.

            End of (7) loop

10)   Set $\mathbf{f}^{(i)}$ equal to whichever $\mathbf{f}_{\beta}^{(i)}$ has the best value of a local performance criterion (to be discussed later). This is the inclusion flag vector that has the best tradeoff between minimizing intra-class separation and maximizing inter-class separation.

    End of (4) loop

End of (2) loop

Here are a few comments on the algorithm just shown:

- Each pass through the iteration loop (2) after the first requires the same amount of time, so runtime is strongly related to the number of iterations. Convergence to stable results is usually very fast, so to keep runtime down, we should use as few iterations as possible.

- In (3) we copy the N by M matrix of binary inclusion flags to a "prior value" work area. This is necessary because when we compute the weights in (5), we will need the entire inclusion flag matrix, but in (10) we will be updating the current inclusion flag matrix **f**. We can't be messing around with the same flag matrix that we are using to compute the weights. So we need a "stable" copy of **f** that we can use for weight computation, and the best way to get this is to just use the flag matrix from the prior iteration. Note that each case i has (while it is being processed as a step of loop (4), but not permanently) its own N-vector of weights for the other cases in the dataset.

- The loop in (4) does not have to be a loop. The computation for each current case (i) is independent of that for all other cases, so all N of these operations can be done simultaneously, in parallel.

- Early in this discussion we brought up the fact that there is a tradeoff involved in minimizing the intra-class separation while maximizing the inter-class separation. We can't do both perfectly, so we need a tradeoff. The loop in (7) finds the optimal tradeoff. In (6) we computed the maximum possible inter-class separation. So to optimize the tradeoff, we place various floors under the inter-class separation, and for each, we compute the minimum intra-class separation. In (10) we compute a performance criterion associated with each trial beta and choose whichever performs the best.

- Unfortunately, the minimization of the intra-class separation in (8) produces real values for the inclusion flags, and we obviously need binary values (a variable either *is* or *is not* included). So we use a random process to convert the real flags to binary flags in a way that should be nearly optimal.

Intra-Class and Inter-Class Separation

We already saw in Equation (3.1) on Page 45 how we can compute the distance between two cases, i and j. We will now make three changes to this distance measure:

1) For mathematical reasons a bit deeper than we can rigorously discuss here, instead of dealing with the Euclidean distance shown in that formula, we will use the square of that distance. In other words, we just omit the square-root operation.

2) We need to incorporate the binary flag vector into the formula, so that we sum squared differences for only those variables whose flag is one and we ignore those variables whose flag is zero.

3) We want to incorporate case weighting into the definition of distance, so that when we compute mean or total distance, we can give varied relative importance to cases.

The symbol $\otimes$ is an operator combining two vectors that means we take the element-by-element product of the two vectors. So our redefined "distance" (it's debatable whether we can still call it a distance, after these three changes!) can be expressed as shown in Equation (3.5). In this equation, i is our "current" case because the inclusion flag vector is that for case i, and the N-component weight vector has been computed in the context of case i.

$$d_{ij} = w_j^{(i)} \left\| \left(x^{(i)} - x^{(j)} \right) \otimes \mathbf{f}^{(i)} \right\|^2 \tag{3.5}$$

Recall from the algorithm shown on Page 58 that we process one case (i) at a time, independently. So we can simplify the notation a lot if we just drop the superscript (i) whenever possible and use j to indicate the "other" case in a distance measurement. All distances are relative to case i, the one we are currently processing, with i omitted as much as possible.

We can also break up vectors using $m=1, ..., M$ to indicate which variable we are talking about. Thus, w_j is the weight associated with case j in the context of the current case i, and f_m is the binary inclusion flag (1 or 0) for variable m in the context of processing case i. Finally, define δ_{jm} as the difference in variable m for cases i and j, as shown in Equation (3.6), and define Δ_j to be the M-vector of the squares of the individual δ components: $\Delta_j = \{\delta_{j1}^2, \delta_{j2}^2 ..., \delta_{jM}^2\}$.

$$\delta_{jm} = x_m^{(i)} - x_m^{(j)} \tag{3.6}$$

We can now write our revised distance measure using individual variables, as shown in Equation (3.7), keeping in mind that everything in this equation is in the context of processing case i so we don't have the burden of (i) superscripts all over the place. I'll even go so far as to drop the i from the subscript in the "distance" measure, so that d_j is the (revised) distance separating our current case i from some other case j. Note that f_m is a vector of ones and zeros, so elementwise squaring does not change anything in it.

$$d_j = w_j \sum_{m=1}^{M} \left(\delta_{jm} f_m \right)^2 = w_j \sum_{m=1}^{M} \delta_{jm}^2 f_m = w_j \Delta_j \mathbf{f} \tag{3.7}$$

At long last we can define intra-class and inter-class separation. In the intuitive development, we talked about the mean separation, but we can just as well talk about total separation, because the mean and the sum differ only by a constant scale factor of the number of cases. Thus, the total intra-class separation is given by Equation (3.8), and the total inter-class separation is given by Equation (3.9).

$$IntraClass = \sum_{j \in class\ of\ i} d_j = \sum_{j \in class\ of\ i} w_j \Delta_j \mathbf{f} \tag{3.8}$$

$$InterClass = \sum_{j \notin class\ of\ i} d_j = \sum_{j \notin class\ of\ i} w_j \Delta_j \mathbf{f} \tag{3.9}$$

The dot product of Δ times $\mathbf{f}$ is associative, so if we define $\mathbf{a}$ and $\mathbf{b}$ as shown in Equations (3.10) and (3.11), respectively, we can define the intra-class and inter-class separation as simple dot products, as shown in Equations (3.12) and (3.13), respectively.

$$\mathbf{a} = \sum_{j \in class\ of\ i} w_j \Delta_j \tag{3.10}$$

$$\mathbf{b} = \sum_{j \notin class\ of\ i} w_j \Delta_j \tag{3.11}$$

$$IntraClass = \mathbf{af} \tag{3.12}$$

$$InterClass = \mathbf{bf} \tag{3.13}$$

To make sure these equations are clear, here is a code snippet taken from LFS_DO_CASE.CPP. We will need a few items. We are processing case which_i in the dataset, the current case. The code is multithreaded, so each thread needs its own work area for **a** and **b**. The following line of code gets the class of the case being processed (whose index has been purposely omitted from the prior few equations to keep the notation simpler). Then we get pointers to this thread's work area for **a** and **b**. Note that M in the equations is n_vars in the code.

```
this_class = class_id[which_i] ;
aa_ptr = aa + ithread * n_vars ;
bb_ptr = bb + ithread * n_vars ;
```

We then compute δ_{jm} using Equation (3.6) for all cases j and variables m. Again, each thread needs its own N by M (n_cases by n_vars) delta matrix.

```
dptr1 = cases + which_i * n_vars ;        // Point to this case
for (j=0 ; j<n_cases ; j++) {
   dptr2 = cases + j * n_vars ;           // Point to case j
   delta_ptr = delta + ithread * n_cases * n_vars + j * n_vars ; // Point to delta for case j
   for (ivar=0 ; ivar<n_vars ; ivar++)
      delta_ptr[ivar] = dptr1[ivar] - dptr2[ivar] ;
   }
```

We can now compute the **a** and **b** vectors. Actually, for a reason that will become clear later, we compute and store the negative of **a**, because that is what we will need. Recall that Δ in Equations (3.10) and (3.11) contains the squares of the δ components.

Also, note that it makes no sense to include j==which_i in this sum, because it's pointless to include the distance between a case and itself in the intra-class separation. But delta for all variables will be zero when j==which_i, so we don't have to skip it in the summation. It's trivially cheaper to go ahead and pointlessly compute it than it would be to check for this condition every time through the loop. Yet some would say that it's cleaner coding to explicitly check for this condition. I include the check here, even though it's really not needed.

```
weights_ptr = weights + ithread * n_cases ;        // This thread's weight vector

for (j=0 ; j<n_vars ; j++)
   aa_ptr[j] = bb_ptr[j] = 0.0 ;

for (j=0 ; j<n_cases ; j++) {
   if (j == which_i)      // Not necessary, because when j==which_i, delta will be zero.
      continue ;          // Still, it's cleaner coding to explicitly check.
```

```
delta_ptr = delta + ithread * n_cases * n_vars + j * n_vars ; // Point to delta for case j
wt = weights_ptr[j] ;
if (class_id[j] == this_class) {
  for (ivar=0 ; ivar<n_vars ; ivar++) { // Equation (3.10)
    term = delta_ptr[ivar] ;
    aa_ptr[ivar] -= wt * term * term ;
    }
  }
else {
  for (ivar=0 ; ivar<n_vars ; ivar++) { // Equation (3.11)
    term = delta_ptr[ivar] ;
    bb_ptr[ivar] += wt * term * term ;
    }
  }
} // For j
```

Computing the Weights

For each "current" case i, we need to compute the N-vector of case weights for all other cases. (We will see that the $j=i$ case weight will never be used, but it's faster and easier to compute it anyway, rather than inserting logic to skip its computation.) We have three primary considerations in weight computation:

1) If case j is far from the current case, its weight should be small, while if it is close, its weight should be large. This focuses attention on the neighborhood of the current case.

2) The neighborhood weighting scheme should not penalize cases that are far simply because they are in different classes. In other words, the weight computation should be relative, not absolute. For example, suppose case j is not in the class of case i, but among all cases not in the current case's class, it is one of the closest. Then its weight should be large, even though on an absolute basis it is far from case i. It may be far away, but relative to other cases not in the current case's class, it is close. If we didn't do this, we would end up largely ignoring cases in classes other than the class of the current case.

3) We don't yet know which metric space (variables selected to define distances) will ultimately be best for defining neighborhoods, so we have to do the best we can. The approach taken in the cited paper and used here is to look at all metric spaces that were computed in the prior iteration and average the weights that would be computed from them. This will cause our weight computation to focus on variables that have been deemed relatively important, while devaluing those variables that have been deemed unimportant.

Up until this point, the distance between the current case i and some other case j has been defined in terms of the metric space of case i, which we notate as $\mathbf{f}^{(i)}$. But for weight computation, we need to be able to discuss the distance between cases i and j in terms of the metric space defined by a third case, k. This is expressed in Equation (3.14).

$$d_{ij|k} = \left\| \left(x^{(i)} - x^{(j)} \right) \otimes \mathbf{f}^{(k)} \right\| \tag{3.14}$$

Above, we observed in the second point that all distances should be relative. In order to accomplish this, we need to know the minimum distance separating the current case i from all other cases that are in the same class as case i. We also need this minimum distance for those cases that are in a different class. And we need this minimum for every metric space k from the prior iteration. These two quantities are expressed in Equations (3.15) and (3.16). In Equation (3.15) we of course must exclude case $j=i$ from the minimum search, because the distance of a case from itself would be zero, which would mean that *MinSame* would be zero all of the time!

$$MinSame_k = \min_{j \in class\, of\, i} d_{ij|k} \tag{3.15}$$

$$MinDifferent_k = \min_{j \notin class\, of\, i} d_{ij|k} \tag{3.16}$$

Once we have these two values computed for a given current case i and metric space k, we can define $dmin_{ij|k}$ as $MinSame_k$ if cases i and j are in the same class, and $MinDifferent_k$ if cases i and j are in different classes. We are now in a position to define the weight for case j (under the assumption that our current case is i). The weight is the average weighting across all metric spaces, and this weighting in any metric space is a negative exponential of the difference between the distance and the minimum same/ different distance for that metric space. This is expressed in Equation (3.17).

$$w_j^{(i)} = \frac{1}{N}\sum_{k=1}^{N}\exp\left(-\left[d_{ij|k} - dmin_{ij|k}\right]\right)$$ (3.17)

This equation is the average over each metric space k of the negative exponential. The term inside the square brackets is the distance between the current case i and the other case j relative to the appropriate minimum distance, according to whether cases i and j are in the same or different classes.

Here is code from the file LFS_WEIGHTS.CPP that demonstrates these computations. As before, because this code is multithreaded, we need to let each thread have its own memory space:

```
wt_ptr = weights + ithread * n_cases ;
d_ijk_ptr = d_ijk + ithread * n_cases ;
```

We begin by zeroing out all of the weights. Then we add in the terms one metric space (k) at a time. Also take note of the class id of the current case.

```
this_class = class_id[which_i] ;

for (j=0 ; j<n_cases ; j++)
  wt_ptr[j] = 0.0 ;
```

This is the outer loop that does one metric space k at a time. The binary variable usage flags, which define each metric space, are in f_prior from the prior iteration. We'll compute d_ijk for all j with fixed which_i and k. While we're at it, we examine all elements of d_ijk and compute two minimums across all j: (1) those for which the class of j is the same as that of which_i, and (2) those in a different class. Recall that we presented the code for computing delta on Page 61. This N by M matrix contains the difference between case i (the current case) and case j (the other case), as expressed in Equation (3.6) on Page 61.

```
for (k=0 ; k<n_cases ; k++) {      // Summation loop builds all weights one k at a time
  fk_ptr = f_prior + k * n_vars ;   // Point to f(k) from the prior iteration
  min_same = min_different = 1.e60 ;
```

```
  for (j=0 ; j<n_cases ; j++) {
    delta_ptr = delta + ithread * n_cases * n_vars + j * n_vars ; // x(i) - x(j)
    sum = 0.0 ;
    for (ivar=0 ; ivar<n_vars ; ivar++) { // Compute norm per Equation (3.14)
      if (fk_ptr[ivar])                    // We're using metric space k
        sum += delta_ptr[ivar] * delta_ptr[ivar] ; // Cumulate (squared) norm
      }
    term = sqrt ( sum ) ;   // This is the norm
    d_ijk_ptr[j] = term ;    // Save it to use soon
    if (class_id[j] == this_class) {   // Equation (3.15)
      if (term < min_same && j != which_i)   // Don't count distance to itself!
        min_same = term ;
      }
    else {                             // Equation (3.16)
      if (term < min_different)
        min_different = term ;
      }
    } // For j, computing d_ijk and the two mins
```

We now have everything we need to compute a term in the sum over k. Cumulate this sum for every weight. Note that we will never use weight[which_i], so we don't need to compute it, but doing so is faster than checking for j==which_i.

```
  for (j=0 ; j<n_cases ; j++) {    // For every weight, add in this k term (Equation (3.17))
    if (class_id[j] == this_class)
      term = d_ijk_ptr[j] - min_same ;
    else
      term = d_ijk_ptr[j] - min_different ;
    wt_ptr[j] += exp ( -term ) ;
    }
  } // For k

 // The sum over k is computed. Divide by N to get average.

 for (j=0 ; j<n_cases ; j++)   // For every weight, add in this k term
   wt_ptr[j] /= n_cases ;
}
```

This section has implemented step 5 in the general algorithm shown on Page 58.

Maximizing Inter-Class Separation

In real-life applications, we often have to deal with the following situation: we need to simultaneously minimize one function while maximizing another, and the two are in conflict. Most of the time, it will be impossible to find a single solution that is located at a minimum of the first function and a maximum of the second. Suppose we have some global criterion that measures the quality of whatever compromise we can obtain. If we are very, very lucky, it will be possible to simply maximize this joint criterion in order to get the solution we want. We are rarely that lucky.

Where we catch a break is if the two optimization problems have a relatively easy solution, including under constraint. In this situation, there is a standard solution to the overall problem, often called the epsilon-constraint method. This algorithm executes in separate steps:

1) Solve the maximization problem and take note of the value of the function obtained at the maximum. Call this epsilon (ε).

2) For a variety of beta (β) values in the range 0 to 1, solve the minimization problem, subject to the constraint that the value of the maximized function at the minimization solution is greater than or equal to $\beta\varepsilon$.

3) Choose whichever solution maximizes the global criterion.

Because the maximized function was able to obtain a value of ε, we know that even if $\beta=1$, we will have a feasible solution to the minimization problem, even though that solution may be far from the minimum. The β value serves as a sort of compromise controller. When $\beta=1$, the maximized function takes full control, while when $\beta=0$, the minimized function is in control. Intermediate values of β provide intermediate importance-weighting of the two conflicting optimization problems.

This is exactly the situation we have with the LFS algorithm. For each case, individually and independently, we want to find a metric space (set of predictor candidates) for that case which minimizes the intra-class separation (Equation (3.12) on Page 61) and also maximizes the inter-class separation (Equation (3.13)). It won't happen often that the same set of predictor candidates will provide optimal solutions to both problems. So we proceed with the epsilon-constraint algorithm as described above:

1) Find the set of predictor candidates that maximizes the inter-class separation, and note the value of this quantity, Equation (3.13), that we are able to obtain, calling this value ε. This is all we need from this step; the identities of these optimal predictor candidates are irrelevant moving forward.

2) For a variety of beta (β) values in the range 0 to 1, minimize the intra-class separation (Equation (3.12)), subject to the constraint that the value of the inter-class separation for the intra-class-optimal set of predictor candidates is greater than or equal to βε.

3) Choose the predictor set corresponding to the β that maximizes a global performance criterion (described later).

It must be pointed out that there is one small fly in this soup. Our predictor flags **f** are binary; a variable is either selected for the metric space or it is not. (Inquiring minds wonder if partial selection might be an interesting subject for study.) However, the optimization routines are strictly real valued; they will find optimal values for the components of **f** that are in the range 0–1 rather than being binary. Later we will see how to deal with this issue. For now, we ignore it.

Notice that both of the functions that we want to maximize/minimize (Equations (3.12) and (3.13)) are linear. Also notice that all of the constraints that we need to impose are linear. For both optimizations, we have the following constraints:

1) All elements of **f** are >= 0.

2) All elements of **f** are <= 1.

3) The sum of the elements of **f** is less than or equal to some user-specified maximum number of variables to use at once.

4) The sum of the elements of **f** is at least 1, meaning that we must select at least one variable in order to have a metric space.

When we minimize the intra-class separation, we will have one additional constraint: the value of the inter-class separation, **bf** per Equation (3.13), must be at least βε.

Since everything is linear, we can do the optimizations using the simplex method. Now is a good time to review the material on interfacing to the Simplex class, which starts on Page 52. The next two lines of code show how to create the Simplex objects, first the one for maximizing the inter-class separation and then the one for minimizing the intra-class separation. Each thread needs its own private object. Let's count constraints, looking at the four categories from above. Category 1 is taken care of automatically, so we can ignore it. Category 2 introduces n_vars constraints. Categories 3 and 4 each introduce a single constraint, making a total of n_vars+2, of which all but the last are of the <= variety. These constraints are used for both optimizations, while the intra-class minimization requires one additional >= constraint.

```
for (ithread=0 ; ithread<max_threads ; ithread++) {
  simplex1[ithread] = new Simplex ( n_vars , n_vars+2 , n_vars+1 , 0 ) ;
  simplex2[ithread] = new Simplex ( n_vars , n_vars+3 , n_vars+1 , 0 ) ;
  }
```

Those creations are done in the constructor. Because most of these constraints are universal, we might as well set them up in the constructor as well. We allocate a work area to hold all constraints for all threads. In this line, nv equals n_vars.

```
constraints = (double *) MALLOC ( (nv+3) * (nv+1) * max_threads * sizeof(double) ) ;
```

We begin by imposing the constraints that no element of **f** can exceed 1. If this code is not clear, please review the Simplex interface.

```
for (i=0 ; i<n_vars ; i++) {                  // For each f upper limit
  constr_ptr = constraints + i * (n_vars+1) ; // Limit (RHS) is first in each row
  constr_ptr[0] = 1.0 ;                        // RHS limit
  for (j=0 ; j<n_vars ; j++)                   // Coefficient of each f
    constr_ptr[j+1] = (i == j) ? 1.0 : 0.0 ;
  }
```

Next, we impose the constraint that the sum of the elements of **f** cannot exceed a user-specified upper limit max_kept on the number of variables that can be selected at once:

```
constr_ptr = constraints + n_vars * (n_vars+1) ;
constr_ptr[0] = max_kept ;
for (j=0 ; j<n_vars ; j++)   // Coefficient of each f
  constr_ptr[j+1] = 1.0 ;
```

Finally, we impose the limit that the sum of the elements of **f** must be at least 1, so that we require at least one variable be selected to define a metric space:

```
constr_ptr = constraints + (n_vars+1) * (n_vars+1) ;
constr_ptr[0] = 1.0 ;
for (j=0 ; j<n_vars ; j++)    // Coefficient of each f
  constr_ptr[j+1] = 1.0 ;
```

The code just seen builds a single constraint matrix. We will need one for each thread, so copy that one to the other threads' constraint matrices.

```
for (ithread=1 ; ithread<max_threads ; ithread++ ) {
  constr_ptr = constraints + ithread * (nv+3) * (nv+1) ;
  for (i=0 ; i<(nv+2)*(nv+1) ; i++)
    constr_ptr[i] = constraints[i] ;
}
```

We still have not dealt with the one additional constraint for minimizing the intra-class separation. We'll take care of that when we need it, as it's easier to follow that way. At least as of now we have taken care of all constraints for the inter-class maximization and all but one for the intra-class minimization.

All that code occurred in the constructor. The actual optimization occurs in the file LFS_DO_CASE.CPP, which handles processing a single case. We already computed **b** (see the code following Equation (3.13) on Page 61). Get a pointer to this thread's copy of it and the constraint matrix:

```
bb_ptr = bb + ithread * n_vars ;
constr_ptr = constraints + ithread * (n_vars+3) * (n_vars+1) ;
```

Then we set the constraints and the objective function, perform the optimization, and get the information about the optimum:

```
simplex1[ithread]->set_objective ( bb_ptr ) ;
simplex1[ithread]->set_constraints ( constr_ptr ) ;
simplex1[ithread]->solve ( 10*n_vars+1000 , 1.e-8 ) ;
simplex1[ithread]->get_optimal_values ( &eps_max , f_real + which_i * n_vars ) ;
```

All we need from the call to get_optimal_values() is the value of the objective function at the maximum, eps_max here and ε in the earlier mathematical discussion. As pointed out in that discussion, we have no need for this optimal **f**. We do need a place for the subroutine to put it, so we send it to the real-valued **f** vector f_real, knowing that it will be overwritten soon.

The parameters of 10∗n_vars+1000 and 1.e-8 in the call to solve() are somewhat arbitrary. Even if there are many thousands of variables, it is unlikely that the number of iterations could exceed this iteration limit. The only purpose of this iteration limit is to be a cheap insurance policy in case the simplex algorithm gets stuck in an endless loop, an occurrence that is theoretically possible but extremely rare in most types of problems. Also, if you have an extremely large problem you might want to relax the 1.e-8 a little to allow for more flexibility in handling accumulated floating-point errors, at the price of slightly inferior optimization.

The algorithm and code presented in this section implemented step 6 in the general algorithm shown on Page 58.

Minimizing Intra-Class Separation

Now that we've found the maximum possible inter-class separation, we can advance to the second phase of the operation, minimizing the intra-class separation for various trial values of β, subject to the additional constraint that the inter-class separation be at least $\beta\varepsilon$. We will need two "best" work areas, so get pointers to thread-private work areas that were reserved in the constructor:

```
best_binary_ptr = best_binary + ithread * n_vars ;
best_fbin_ptr = best_fbin + ithread * n_vars ;
```

The constraint matrix created in the constructor (Page 69) is complete except for the last constraint, which is that the inter-class separation (Equation (3.13) on Page 61) be at least $\beta\varepsilon$. Insert this constraint now. Note that the coefficients of this constraint, **b**, will be the same for all trial values of β, but the limit, $\beta\varepsilon$, will of course change for each trial. So set all of the coefficients now, and leave the limit for the actual testing of a trial value.

```
temp_ptr = constr_ptr + (n_vars+2) * (n_vars+1) ; // Last row in constraint matrix
for (j=0 ; j<n_vars ; j++) // Coefficient of each f
   temp_ptr[j+1] = bb_ptr[j] ;
```

The following loop, step 7 in the algorithm shown on Page 58, tests as many equally spaced trial values of β as the user specifies (n_beta). The routine test_beta() returns the negative of the intra-class separation in crit. This is because the simplex algorithm maximizes, so if we want to minimize, we must maximize the negative of our function. This is also why we flipped the sign of **a** when we computed it, as discussed starting on Page 62.

```
best_crit = -1.e60 ;
for (i=1 ; i<=n_beta ; i++) { // Trial values of beta
  test_beta ( which_i , (double) i / (n_beta+1) , eps_max , &crit , ithread ) ;
  if (crit > best_crit) {
    best_crit = crit ;
    for (ivar=0 ; ivar<n_vars ; ivar++)              // Copy optimal f for this beta
      best_fbin_ptr[ivar] = best_binary_ptr[ivar] ; // to best f across all betas
    }
  }
```

The routine test_beta() (discussed in the next section) also sets the private array pointed to by best_binary_pointer to the binary version of **f** that corresponds to the trial β. In the code just seen, we test the value of crit just returned against the best so far. If we set a new record for a trial β, we copy the binary **f** to a local copy of the best, best_fbin_ptr. When we are finished testing numerous values, the best **f** is there, so we save it in its final area:

```
iptr = f_binary + which_i * n_vars ;
for (ivar=0 ; ivar<n_vars ; ivar++)
  iptr[ivar] = best_fbin_ptr[ivar] ;
```

It is important to understand that f_binary does not exist in different versions for different threads. This is a single N by M matrix in which row i contains the M inclusion flags for case i. There will be no contention for elements of this array by different threads. Each thread will compute a single row of this array, with this row being which_i for the thread. Completion of this array is the ultimate goal of the entire LFS algorithm, and we are free to compute these rows sequentially, or all at once in parallel, or with any sort of multithreading we wish. Each row is computed separately and independently, the product of a single which_i. The three lines of code just shown complete step 10 of the general algorithm shown on Page 58.

Testing a Trial Beta

The routine test_beta(), implemented in the file LFS_BETA.CPP, is given a trial value of β. It then does three things:

1) It minimizes the intra-class separation, subject to the extra constraint that the inter-class separation for the "optimal" **f** vector is at least $\beta\varepsilon$. This is step 8 in the general algorithm shown on Page 58.

2) It uses a random process to convert the real-valued **f** just computed to a binary **f**.

3) It evaluates the quality of the binary **f** so that we can keep track of which trial β is best and thereby choose the best **f** for this case i.

The first step is to set the $\beta\varepsilon$ constraint. Recall that the coefficients of this constraint are the same for all trial betas, so they were set in the caller (Page 71). But we do have to set the limit here, because it changes for each trial beta. And then we just set the objective function and constraints, optimize, and get the (real, not binary) **f** vector. We don't need the value of the intra-class separation here; all we need is the optimal **f**.

```
constr_ptr[(n_vars+2)*(n_vars+1)] = beta * eps_max ;

simplex2[ithread]->set_objective ( aa_ptr ) ;
simplex2[ithread]->set_constraints ( constr_ptr ) ;
simplex2[ithread]->solve ( 10 * n_vars + 1000 , 1.e-8 ) ;
fr_ptr = f_real + which_i * n_vars ;
simplex2[ithread]->get_optimal_values ( &dtemp , fr_ptr ) ;
```

We have real values for an **f** that is optimal, but we need binary values for **f**. So we use a Monte-Carlo method to find the best binary vector. To do this, repeat a simple process n_rand times, where n_rand is a user-specified parameter generally in the range of 500 to several thousand, depending on the number of variables. Each time, we do the following:

> For each variable, set its binary flag to 1 with probability equal to the real value of **f** for that variable. If the binary vector satisfies the constraints of the simplex minimization, compute its value for the objective function and keep track of the best function value across all n_rand tries. Whichever such randomly generated binary vector has the minimum objective function is chosen as the binary **f** vector.

We don't have to worry about this looping insanely long due to repeated failure to satisfy the constraints, because the only relevant constraint is the number of "1" flags, and that is satisfied with high-enough probability to avoid massive failures. We don't bother checking for satisfaction of the **bf**>=βε constraint, because doing that could tremendously slow things. In fact, it is conceivable that because ε was determined using real **f** values, there may be *no* binary **f** that satisfies that constraint, in which case we would keep looking forever!

To initialize, we get a pointer to the "master" array of binary inclusion flags, although at this point we are just using it as a scratch vector. It is set to its final value at the end of LFS_DO_CASE.CPP, which was discussed on Page 72. In truth, we are minimizing the intra-class separation, but recall that our objective function coefficients are really –**a** so in the code we maximize. Set the best value (the maximum so far) to a very negative number. Finally, initialize the seed for the random number generator. Then loop.

```
fb_ptr = f_binary + which_i * n_vars ;   // Point to the binary vector for this which_i
best_func = -1.e60 ;
iseed = which_i + 1 ;

for (irand=0 ; irand<n_rand ; irand++) {
  n = 0 ;
  for (ivar=0 ; ivar<n_vars ; ivar++) {
    if (fast_unif(&iseed) < fr_ptr[ivar]) { // Set binary to 1 with this probability
      fb_ptr[ivar] = 1 ;
      ++n ;               // Count this '1'
      }
    else
      fb_ptr[ivar] = 0 ;
    }

  if (n == 0 || n > max_kept) {     // The only relevant constraint
    --irand ;                       // If we fail, just try again
    continue ; // This should not happen very often, so no worries
    }
```

In the preceding loop, n counts the number of 1's in the trial binary **f**. The two constraints that we check are that n is at least 1 (so that at least one variable is included) and it does not exceed the user-specified maximum number of variables in a metric space. If either constraint is not satisfied, we just ignore that trial and try again. We need not worry about this becoming an overly long loop, because constraint failures should be relatively uncommon. After all, the real values that are being used as probabilities satisfy these constraints.

At this point, fb_ptr contains a trial binary vector that satisfies the constraints on n. Evaluate its (negative) intra-class separation and keep track of the best so far in best_binary_ptr.

```
sum = 0.0 ;
for (ivar=0 ; ivar<n_vars ; ivar++)
  sum += aa_ptr[ivar] * fb_ptr[ivar] ;

if (sum > best_func) {
  best_func = sum ;
  for (ivar=0 ; ivar<n_vars ; ivar++)
    best_binary_ptr[ivar] = fb_ptr[ivar] ;
  }
}
```

When we finish the loop just shown, best_binary_ptr contains the binary **f** that was the best (minimum intra-class separation) among all random tries that satisfied the constraints on the number of variables included. We're almost done. All we need to do is evaluate a measure of the quality of this binary **f** in terms of its ability to be a good local classifier. Remember the goal of this step in the algorithm (step 10 in the general algorithm shown on Page 58); this routine evaluates the quality of a trial beta.

Here is the only place in this presentation in which I depart from the algorithm described in the paper cited at the beginning of this chapter. Their method for evaluating quality is almost certainly very good, but it is quite complex and slow to compute. I devised a much faster method that I have reason to believe is nearly or equally as good, maybe better. ☺

My method is vaguely related to the Mann–Whitney U test. Here is the philosophy behind my test. If this **f** (which was entirely computed in the context of case i, or which_i in the code) is good and we examine cases near x_i per the metric space defined by **f**, those cases that are in the class of case i should be closer to x_i than those not in that class. So we compute these distances and use a measure of how well this ordering occurs. We borrow d_ijk as a work area for the distances. The following code computes these distances. We also initialize an index array nc_iwork_ptr. Recall that delta, the difference between each variable for each case, was computed as shown on Page 62.

```
for (j=0 ; j<n_cases ; j++) {
  delta_ptr = delta + ithread * n_cases * n_vars + j * n_vars ; // Point to delta for case j
  sum = 0.0 ;
  for (ivar=0 ; ivar<n_vars ; ivar++) {
    if (best_binary_ptr[ivar])   // The distance is defined by this metric space
      sum += delta_ptr[ivar] * delta_ptr[ivar] ;
    }
  d_ijk_ptr[j] = sum ;  // Distance of case j from this (which_i) case in f space
  nc_iwork_ptr[j] = j ; // We'll need this index soon
  }
```

Now we sort these distances in ascending order and convert them to ranks, with the closest having the lowest rank. We simultaneously move the indices of the cases so that later we can identify a case from its position in the sorted array.

```
qsortdsi ( 0 , n_cases-1 , d_ijk_ptr , nc_iwork_ptr ) ; // Sort ascending, moving index

for (j=0 ; j<n_cases ; ) {
  val = d_ijk_ptr[j] ;
  for (k=j+1 ; k<n ; k++) { // Find all ties
    if (d_ijk_ptr[k] > val)
      break ;
    }
  rank = 0.5 * ((double) j + (double) k + 1.0) ;
  while (j < k)
    d_ijk_ptr[j++] = rank ;
  } // For each case in sorted distance array
```

We now have the ranks of the distances in d_ijk_ptr. The final step is to compute the performance criterion, which will be returned to the caller in crit. The method is to pass through all cases to cumulate a total score. If a case is in the class of case i, we subtract its rank, weighted by that case's weight already computed and used for prior operations. If the case is in a different class, we add the weighted rank. Suppose that this **f** is very high quality. Then we will be subtracting small numbers (close distances have low ranks) for cases in the class of case i and adding large numbers (large distances have large ranks) for cases in a different class. Thus, our criterion will be strongly related to how well this metric space distinguishes classes. Moreover, these plus or minus contributions to the criterion are weighted according to the locality of the cases relative to case i, favoring local performance.

```
*crit = 0.0 ;
for (j=0 ; j<n_cases ; j++) {
   k = nc_iwork_ptr[j] ;      // Original index of this sorted case
   if (k == which_i)          // No sense scoring a case with itself!
      continue ;
   if (class_id[k] == this_class)
      *crit -= d_ijk_ptr[j] * weight_ptr[k] ;
   else
      *crit += d_ijk_ptr[j] * weight_ptr[k] ;
   }
```

It must be mentioned that a recent paper presents an alternative method for converting the real-valued **f** to a binary **f**: "A Fast Algorithm for Local Feature Selection in Data Classification," by Fereshteh Sadat Hoseininejad, Yahya Forghani, and Omid Ehsani, to appear in *Expert Systems* at a date unknown at the time of this writing. Their method provides a closed-form solution that is considerably faster to compute than the random method shown here. When there are relatively few cases, this can make an enormous difference in computation time. However, if there are several thousand cases, this binary conversion takes up less than one percent of total compute time, and so this new algorithm will make no perceptible difference. Because my own work always involves a large number of cases, I have not pursued this new algorithm.

A Quick Note on Threads

Multithreading for the LFS algorithm is particularly easy because each thread is a single case. There is no interaction between threads and no fancy parameter passing. The multithreading is done in the run() routine found in LFS.CPP. A detailed description of multithreading can be found starting on Page 18 in the context of Forward Selection Component Analysis. I have every confidence that if the reader studies that presentation, the much simpler LFS code will be self-explanatory.

CUDA Computation of Weights

This section has three prerequisites, and if the reader does not satisfy them all, there is little point in continuing (other than to satisfy curiosity, which is a noble task). These are as follows:

1) The algorithm for computing weights described starting on Page 63 will be implemented in exactly the same way mathematically, but in a radically different way as far as computation is concerned. This is because CUDA computation is inherently parallel, while the algorithm already shown is serial. If you're going to understand the parallel version, you must completely understand the serial version and the underlying mathematics.

2) The CUDA version of weight computation occurs in several steps that are separated in the general algorithm shown on Page 58, as opposed to being computed as a single step (5) in the serial version. So you need to fully understand that overview of the algorithm.

3) You need a basic working understanding of CUDA programming. You do *not* need to be an expert, as this code is straightforward, but you do need to be able to write simple CUDA code. This section will guide you in how to write weight computation CUDA code, but it is not a CUDA tutorial.

All of the CUDA code found in this section is in LFS_CUDA.cu, which includes most error checking code. The listings shown and discussed here have error checking code omitted for clarity. A responsible programmer will check the return value of every CUDA routine that can possibly fail and take appropriate steps in case of failure.

Integrating the CUDA Code into the Algorithm

Each CUDA module will be described in its own subsection, but first we examine an overview of how these modules relate to the general algorithm shown on Page 58.

Before anything else is done, we must copy the dataset of predictor candidates to the CUDA hardware using lfs_cuda_init(). I find that the LFS constructor is a good place to do this. And the LFS destructor is a great place to call lfs_cuda_cleanup() to release all of the hardware memory that was used. In any case, these calls would be before the start and after the end of the overall Page 58 algorithm, respectively.

Between steps 1 and 2, we call lfs_cuda_classes() to copy the class ID of each case to the CUDA hardware. If you are absolutely positive that neither you nor any other user will ever be doing a Monte-Carlo permutation test, you could do this in the constructor immediately after calling lfs_cuda_init(). But it is safer and better to do this just before beginning the main iteration loop. That way, if the class IDs are permuted between multiple invocations of the LFS algorithm, the permuted classes will be copied to the hardware.

Between steps 3 and 4, at the start of the main iteration loop but before processing the cases, we must call lfs_cuda_flags() to copy the prior iteration's binary flags to the CUDA hardware for the weight computation algorithm. This is the CUDA equivalent of step 3 in the general algorithm.

Once we are ready to start processing individual cases (in step 4), we execute a series of CUDA calls that compute the weights. Of course, we don't need to do this on the first step 2 iteration, because the weights are all one for the first pass. If the rest of the LFS computation is multithreaded, the only sensible approach, we do the CUDA weight computation just before launching each thread. This has the lovely bonus of overlapping host CPU computation (processing previously launched threads) with this CUDA weight computation, putting everybody to work at once.

The CUDA routines that are called when we are about to launch a new host thread are as follows:

> lfs_cuda_diff() – Compute the n_vars by n_cases matrix of differences between the current case i and each "other" case j. These are the $x^{(i)} - x^{(j)}$ terms in Equation (3.14) on Page 64.

> lfs_cuda_dist() – Compute the n_cases by n_cases matrix of distances between the current case i and each "other" case j, measured by each of the n_cases metrics. This is the full Equation (3.14) on Page 64. Recall that i is fixed, so the metric k is the row of this matrix and the "other" case j is the column.

> lfs_cuda_mindist() – Compute two n_cases vectors. One is the minimum across all "other" cases of their distance from the current case i, measured by each of the n_cases metrics, with the minimum found for only "other" cases in the same class as the

79

current case i. The other vector is this same thing, except with the minimum computed for only those "other" cases not in the class of case i. These are Equations (3.15) and (3.16), respectively.

lfs_cuda_term() – Compute each of the terms in Equation (3.17), an n_cases by n_cases matrix.

lfs_cuda_transpose() – Transpose the term matrix just computed so that we can use the parallel reduction algorithm to do the summations. Each row of this matrix consists of the terms for a single weight j, and each column is for metric k.

lfs_cuda_sum() – Sum the rows of the transposed term matrix and divide by n_cases to get the weights, as expressed in Equation (3.17).

lfs_cuda_get_weights() – Transfer the just-computed weight vector from the CUDA hardware to the host.

Initializing the CUDA Hardware

Before doing anything else, we must call lfs_cuda_init() to allocate all required memory on the device and copy the dataset to it. This copy operation involves only the predictor candidates, not the class ID vector which associates classes with cases. This is so we can embed LFS runs inside a Monte-Carlo permutation loop that shuffles the class IDs.

For handy reference, here is a listing of the host-static variables and their CUDA hardware counterparts. Names that begin with d_ reside on the device and are used by the hardware routines. Names that begin with h_ reside in the host and equal the device value. This lets us save a little time by avoiding the need to pass a bunch of parameters in the launch. We could just pass pointers as parameters, but that's overhead. So instead we use cudaMemcpyToSymbol() to copy the values on the host to values on the device. This lets global routines address the values that are already set on the device rather than having to use passed parameters.

```
static    int ncases ;           // Number of cases
__constant__ int d_ncases ;
static    int nvars ;            // Number of variables (predictor candidates)
__constant__ int d_nvars;
```

```
static     int ncols ;                   // Number of columns in next five matrices
__constant__ int d_ncols ;               // Which is (ncases+31)/32*32 for memory alignment
static     float *h_data ;               // nvars by ncases data matrix, case changes fastest
__constant__ float *d_data ;
static     float *h_diff ;               // nvars by ncases difference matrix
__constant__ float *d_diff ;
static     float *h_dist ;               // ncases by ncases distance matrix [metric, jcase]
__constant__ float *d_dist ;
static     float *h_trans ;              // ncases by ncases transposed term matrix
__constant__ float *d_trans ;
static     int *h_flags ;                // nvars by ncases binary matrix (fprev)
__constant__ int *d_flags ;
static     int *h_class ;                // ncases class ID vector
__constant__ int *d_class ;
static     float *h_minSame_out ;
__constant__ float *d_minSame_out ;
static     float *h_minDiff_out ;
__constant__ float *d_minDiff_out ;
```

The initialization routine begins as shown in the following text. We copy the number of cases and variables to their static counterparts. One crucial action is to round up the number of cases (which are columns on the device) to a multiple of 128 bytes. This is because the CUDA hardware uses a 128-byte path to access the global data cache. Each global memory access transfers a multiple of 128 bytes, no matter how many bytes are actually requested. By ensuring that every row in the matrices starts on an address that is a multiple of 128 bytes, we ensure that global data accesses are efficient.

Note that cuda_present has previously been set to the number of CUDA devices present on the hardware via a call to cudaGetDeviceCount() (in CUDAHDWR.CPP). The first device (0) is generally the one used for the video display. If we have several CUDA devices, by avoiding the first for computation, we avoid competing with the display. Finally, we get some hardware information and copy dimension values to the device.

```
int lfs_cuda_init (
  int n_cases ,        // Number of cases
  int n_vars ,         // Number of columns in host database (rows on device)
  double *data ,       // Data matrix, n_cases by n_vars
```

```
  char *error_msg      // Error message returned here
  )
{
  int i, j ;
  float *fdata ;
  cudaError_t error_id ;

  ncases = n_cases ;    // These are static in this module
  nvars = n_vars ;       // This is the number of predictor candidates
  ncols = (ncases + 31) / 32 * 32 ; // Bump up to multiple of 128 bytes for alignment

  error_id = cudaSetDevice ( cuda_present - 1 ) ;
  cudaGetDeviceProperties ( &deviceProp , 0 ) ;

  cudaMemcpyToSymbol ( d_ncases , &ncases , sizeof(int) , 0 ,
                            cudaMemcpyHostToDevice ) ;
  cudaMemcpyToSymbol ( d_nvars , &nvars , sizeof(int) , 0 ,
                            cudaMemcpyHostToDevice ) ;
  cudaMemcpyToSymbol ( d_ncols , &ncols , sizeof(int) , 0 ,
                            cudaMemcpyHostToDevice ) ;
```

We'll present the code that allocates memory for the dataset and copies it from the host to the device. This is the only memory copy during initialization, and other allocations are similar, so in the interest of economy, we'll omit them. Here is the code, and a short discussion follows.

```
  memsize = ncols * nvars * sizeof(float) ;

  error_id = cudaMalloc ( (void **) &h_data , (size_t) memsize ) ;
  if (error_id != cudaSuccess) {
    // Handle error reporting
    }

  cudaMemcpyToSymbol ( d_data , &h_data , sizeof(void *) , 0 ,
                            cudaMemcpyHostToDevice ) ;

  // Copy it from host, transposing so cases change fastest

  fdata = (float *) MALLOC ( memsize ) ;
```

```
for (i=0 ; i<ncases ; i++) {
  for (j=0 ; j<nvars ; j++)
    fdata[j*ncols+i] = (float) data[i*nvars+j] ;
  }

error_id = cudaMemcpy ( h_data , fdata , memsize , cudaMemcpyHostToDevice ) ;
FREE ( fdata ) ;
fdata = NULL ;

if (error_id != cudaSuccess) {
  // Handle error reporting
  }
```

We allocate space for a row for each candidate predictor, and the number of columns is ncols rather than n_cases so that memory accesses align for maximum transfer efficiency. We then copy the address of the allocated memory (h_data) to the device (d_data) for fast global access.

The data is stored on the host with variables changing fastest, but for fast memory access, we store it on the device with cases changing fastest. Also, the data is double on the host but float on the device, so we translate and transpose using the temporary work array fdata.

Computing Differences from the Current Case

Equation (3.14) on Page 64 requires us to have the difference between each "other" case j and the current case i. We'll store this on the hardware as a matrix with n_vars rows and n_cases columns. Recall that like all matrices we use for this LFS project, the allocation actually reserves space for ncols columns, where ncols is n_cases rounded up to a multiple of 128 bytes (32 4-byte integers). Therefore, unless there happens to be a multiple of 32 cases, the rows are incomplete. The ncols–n_cases columns at the end of each row are completely ignored and may be left uninitialized.

We could just compute this difference matrix on the host and then copy it to the CUDA device, thereby saving some device memory (that for holding the dataset). But compared to the several much larger matrices we have to store, this is a relatively small amount of memory, and changing the order of computation would significantly complexify the host code. Also, computing this difference matrix is blink-of-an-eye fast for CUDA, much faster than serial computation on the host. So the choice is simple.

Here is the launch code for this routine. We get the warp size, even though it will likely be 32 for our lifetimes. We let the block size be the number of cases, rounded up to an integer number of warps, and then limit it to a reasonable number of warps. Because this algorithm will use no shared memory and have very few registers, we could let the blocks be larger, but in deference to users who may have old hardware, we set the limit to 8 warps. Cases (j) are indexed by the thread and variables by the block's y index. Launch the kernel and then wait for it to end.

```
warpsize = deviceProp.warpSize ;      // Threads per warp, likely 32 well into the future
threads_per_block = (ncases + warpsize - 1) / warpsize * warpsize ;
if (threads_per_block > 8 * warpsize)
   threads_per_block = 8 * warpsize ;

block_launch.x = (ncases + threads_per_block - 1) / threads_per_block ;
block_launch.y = nvars ;
block_launch.z = 1 ;

lfs_cuda_diff_kernel <<< block_launch , threads_per_block >>> ( icase ) ;
cudaThreadSynchronize () ;
```

Here is the kernel code. The index of the "other" case j is procured in the usual way. We need three pointers: that for the current case i, that for the "other" case j, and that for where we place the difference. Each stored row is n_ncols long, even though only the first n_cases elements contain valid data.

Experienced CUDA programmers will immediately spot the correct global data alignment in this routine, but for the sake of beginners, I'll mention it. When the memory for the data and difference matrices was allocated, the compiler guaranteed that each would begin on an address that is a multiple of 128 bytes. Also, blockDim.x (threads_per_block) is guaranteed to be a multiple of 128 bytes by the launch code just shown, and d_ncols is a multiple of 128 bytes by design. Thus, for the first thread in each warp, jcase_ptr and diff_ptr will be at an address that is a multiple of 128 bytes. As a result, the fetches from and stores to global memory for these two arrays will perfectly align with the 128-byte cache line, resulting in optimal data transfer.

Things are a little different but still excellent for icase_ptr. We don't need to think about alignment issues because this address is the same for all threads in the block. Thus, the fetched value is shared, meaning that only a single global fetch is needed.

```
__global__ void lfs_cuda_diff_kernel ( int icase )
{
  int jcase ;
  float *icase_ptr, *jcase_ptr, *diff_ptr ;

  jcase = blockIdx.x * blockDim.x + threadIdx.x ;
  if (jcase >= d_ncases)
    return ;

  icase_ptr = d_data + blockIdx.y * d_ncols + icase ; // which_i
  jcase_ptr = d_data + blockIdx.y * d_ncols + jcase ; // j
  diff_ptr = d_diff + blockIdx.y * d_ncols + jcase ;

  *diff_ptr = *icase_ptr - *jcase_ptr ;
}
```

Computing the Distance Matrix

This routine computes the n_cases square matrix of distances that are defined by
Equation (3.14) on Page 64. Each row of the matrix corresponds to a metric k, and each
column corresponds to a case j. Memory alignment and other efficiency measures are
particularly important, because in large problems, this one routine consumes more
device time than all of the other CUDA routines put together.

The launch code is almost identical to that for the difference routine just presented,
except that the y block dimension is n_cases rather than n_vars.

```
warpsize = deviceProp.warpSize ;    // Threads per warp, likely 32 well into the future

threads_per_block = (ncases + warpsize - 1) / warpsize * warpsize ;
if (threads_per_block > 8 * warpsize)
  threads_per_block = 8 * warpsize ;

block_launch.x = (ncases + threads_per_block - 1) / threads_per_block ;
block_launch.y = ncases ;
block_launch.z = 1 ;

lfs_cuda_dist_kernel <<< block_launch , threads_per_block >>> () ;

cudaThreadSynchronize () ;
```

The device code is shown on the next page. Experienced CUDA programmers will cringe at the sight of the summation loop over all variables, instantly thinking that there must be a better way, a more CUDA-friendly approach. And indeed, this looping may cause a problem with video display timeouts if there are an enormous number of variables, especially if the routine is running on old hardware. On the other hand, on my GTX 1080Ti, with over 2000 cases and 220 variables, a launch runs in just two milliseconds. Thus, I believe that there is a tremendous amount of leeway, and breaking the launch into several launches that do partial sums would be overkill. And with large real-life applications, little or nothing would be gained in terms of granularity, because we are launching n_cases blocks in the y dimension and about n_cases/256 in the x dimension.

```
__global__ void lfs_cuda_dist_kernel ()
{
  int ivar, jdiff, *flags_ptr ;
  float *jdiff_ptr, *dist_ptr, sum ;

  jdiff = blockIdx.x * blockDim.x + threadIdx.x ;
  if (jdiff >= d_ncases)
    return ;

  jdiff_ptr = d_diff + jdiff ;             // First variable for case j
  flags_ptr = d_flags + blockIdx.y ;  // First flag for metric case blockIdx.y

  sum = 0.0f ;
  for (ivar=0 ; ivar<d_nvars ; ivar++) {
    if (*flags_ptr)
      sum += *jdiff_ptr * *jdiff_ptr ;
    jdiff_ptr += d_ncols ;
    flags_ptr += d_ncols ;
    }
  dist_ptr = d_dist + blockIdx.y * d_ncols + jdiff ; // distance[metric,jcase]
  *dist_ptr = sqrt ( sum ) ;
}
```

It's instructive to examine memory accesses in this code. We are reading two global arrays: the difference matrix just computed (in d_diff) and the binary inclusion flag matrix (in d_flags), called **f** in the mathematical presentations. We are writing a single array, the computed distance matrix (d_dist). The alignment issue discussed in the prior section on computing differences holds here as well for d_diff and d_dist. They start out at an address that is a multiple of 128 bytes for the first thread in each warp, and because we are adding d_ncols inside the summation loop, this property continues to hold throughout execution of the launch. Thus, a single 128-byte transaction will service every thread in a warp.

Things are also excellent though a bit different with d_flags. The fetch address is the same for all threads in the block, and hence the fetched value is shared. Thus, only a single fetch is needed.

Finally, note that because d_flags is identical in all threads of a block, there are no branching stalls.

Computing the Minimum Distances

Once we have the matrix of every case's distance from the current case i, measured in every metric k, we have to compute two vectors, each n_cases long. These are the "same class" and "different class" minimums across all "other" cases j, defined by Equations (3.15) and (3.16) on Page 64.

This routine uses a relatively advanced algorithm called *reduction*, which is considerably more complex than the nearly trivial code seen so far. I won't delve into a full tutorial on this important parallel algorithm but rather just summarize the code presented. Readers not familiar with reduction can find it explained in most CUDA programming books. I also have a gentle tutorial on it in Volume 1 of my "Deep Belief Nets in C++ and CUDA C" series.

Two separate kernels are launched to perform this task, with the first being the complex one, shown on the next page, with a discussion following. Here is the launch code for that first algorithm. Note that the two work/output arrays shown here were allocated in lfs_cuda_init().

```
#define REDUC_THREADS 256 /* Must be a power of two! */
#define REDUC_BLOCKS 64

// This allocation was done in the initialization code
   memsize = REDUC_BLOCKS * ncols * sizeof(float) ;
   error_id = cudaMalloc ( (void **) &h_minSame_out , (size_t) memsize );
   error_id = cudaMalloc ( (void **) &h_minDiff_out , (size_t) memsize );
```

```
// Launch code
  blocks_per_grid = (ncases + REDUC_THREADS - 1) / REDUC_THREADS ;
  if (blocks_per_grid > REDUC_BLOCKS)
    blocks_per_grid = REDUC_BLOCKS ;
  orig_blocks_per_grid = blocks_per_grid ;

  block_launch.x = blocks_per_grid ; // Case for distance
  block_launch.y = ncases ;          // Metric
  block_launch.z = 1 ;

  lfs_cuda_mindist_kernel <<< block_launch , REDUC_THREADS >>> ( which_i ) ;
  cudaDeviceSynchronize() ;

__global__ void lfs_cuda_mindist_kernel ( int which_i )
{
  __shared__ float partial_minsame[REDUC_THREADS],
                   partial_mindiff[REDUC_THREADS] ;
  int j, index, iclass ;
  float *dist_ptr, min_same, min_diff ;

  index = threadIdx.x ;
  iclass = d_class[which_i] ;

  dist_ptr = d_dist + blockIdx.y * d_ncols ;   // This metric

  min_same = min_diff = 1.e30 ;
  for (j=blockIdx.x*blockDim.x+index ; j<d_ncases ; j+=blockDim.x*gridDim.x) {
    if (d_class[j] == iclass) {
      if (dist_ptr[j] < min_same && j != which_i)
        min_same = dist_ptr[j] ;
      }
    else {
      if (dist_ptr[j] < min_diff)
        min_diff = dist_ptr[j] ;
      }
    }
```

```
partial_minsame[index] = min_same ;
partial_mindiff[index] = min_diff ;
__syncthreads() ;

for (j=blockDim.x>>1 ; j ; j>>=1) {
  if (index < j) {
    if (partial_minsame[index+j] < partial_minsame[index])
      partial_minsame[index] = partial_minsame[index+j] ;
    if (partial_mindiff[index+j] < partial_mindiff[index])
      partial_mindiff[index] = partial_mindiff[index+j] ;
  }
  __syncthreads() ;
}

if (index == 0) { // min [ sub-part , metric ]
  d_minSame_out[blockIdx.x*d_ncols+blockIdx.y] = partial_minsame[0] ;
  d_minDiff_out[blockIdx.x*d_ncols+blockIdx.y] = partial_mindiff[0] ;
}
}
```

The two arrays whose allocations are shown here (though they were actually allocated during initialization) serve as both scratch storage for the reduction, as well as output of the two vectors defined by Equations (3.15) and (3.16) on Page 64.

Recall that the distance matrix that will be the source data for this algorithm is organized with each row corresponding to a metric. Our goal is to find the minimum (separately for "same class" and "different class") across each row. The discussion that now appears considers an individual row. The y coordinate of the block determines the row being processed, and these rows are processed independently and identically, so there is no interaction between rows. We ignore the row in this discussion.

Visualize a row of distances as a single horizontal vector. Now, in your mind's eye, restructure that vector as a matrix having REDUC_THREADS columns. The first row of this imaginary matrix will contain the first REDUC_THREADS distances per the metric being processed, the next row will contain the next REDUC_THREADS distances, and so forth. The number of rows in this restructured matrix will obviously be n_cases / REDUC_THREADS, rounded up to the next integer (the last row will usually be incomplete). If you review the launch code, you will see that blocks_per_grid, the x dimension of the grid, is set equal to this quantity, except that it is limited to our predefined REDUC_BLOCKS, a quantity that is not terribly critical to speed and has no impact on correctness.

We now work our way through the reduction kernel. Observe that we allocate two shared arrays, one for the "same class" computation and one for the "different class." This is extremely fast on-chip memory that is shared by all threads in the block.

The index is determined strictly by the thread, with the block playing no role. We fetch the class of the current case i, a global memory access that is constant for all threads in the block and hence accomplished with a single transaction. We set dist_ptr to point to the vector of distances corresponding to the metric y.

The minimum same-class and different-class distances are initialized to huge numbers. The loop that follows is, in practice, nearly always pointless in this application, but it is included to be strictly correct (even though it does very slightly slow execution of the kernel). We'll return to this point in a moment, but first look at what happens inside the loop.

For each pass through the loop, we are considering an "other" case j whose distance from the current case i is in dist_ptr[j]. If this case is in the same class as the current case, we update the minimum "same-class" distance, while if it is not in the same class, we update the minimum "different-class" distance. Of course, if they are in the same class, we have to make sure they are not just the same case, because that distance will be zero, which would make the test pointless!

Back to the loop. Consider the first pass, and recall our imaginary matrix visualization of this vector of distances. The first thread in the first block will point to the upper-left element in our imaginary matrix, the first column in the first row. The next thread in this block points to the next column in this row, and so forth, until the maximum value of the thread, REDUC_THREADS−1, points to the last column in this row. Similarly, the first thread in the second x block points to the first column in the second row of our imaginary matrix. In other words, the x coordinate of the block specifies the row of our imaginary matrix, and the thread specifies the column.

How many elements can this imaginary matrix contain? There are blocks_per_grid rows, with this quantity limited to REDUC_BLOCKS=64 (see the launch code), and there are REDUC_THREADS=256 columns. So our imaginary matrix can hold at most 64∗256=16,384 elements.

The first time through this loop, j points to the element in our imaginary matrix that is at row blockIdx.x and column index. After this element is processed (used to update the minimums), we add blocks_per_grid times REDUCT_THREADS to j. If blocks_per_grid has not been limited to REDUC_BLOCKS in the launch, it will have been set to exactly the number of rows needed to contain all of the cases, with the last row usually being incompletely filled. Thus, adding the loop increment will put j past d_ncases, causing the loop to end after the first pass. Because this limiting will not occur unless the number of cases exceeds 16,384, one pass through this loop will be all that happens in any practical application of the *VarSceen* program. Thus, some readers may wish to change this from a loop to a single instruction, thereby saving a small amount of compute time. I wrote it this way to make clear the "correct" implementation of the algorithm.

To be thorough, let's consider what would happen if the number of cases did exceed REDUC_BLOCKS times REDUC_THREADS, even though this would be extremely unlikely in *VarScreen*. In the second pass through the loop, we would be looking at a case in the same column as in the first pass and exactly REDUC_BLOCKS rows lower. The contribution of this element to the same-class and different-class minimums would be accounted for. On a third pass, we would be looking at an element another REDUC_BLOCKS rows lower, and so forth. The bottom line is this: after the loop is complete, any data beyond those initial REDUC_BLOCKS blocks has been accounted for in the two minimums and may henceforth be ignored.

Astute readers may wonder what happens in regard to the unused entries at the end of the last row. For example, recall that REDUC_THREADS has been set to 256. What if there are 257 cases? We will launch two *x* blocks (two rows of our imagined matrix), but the second row will have only one valid case; its remaining 255 columns will be uninitialized garbage. In this situation, only the thread whose index is zero will process the loop. The other 255 threads in the second block will not even make a first pass through the loop. For example, when the index in the second block (blockIdx.x=1) is 1, the starting value of j is 1*256+1=257, which is not less than d_ncases=257. Thus, the loop for this thread 1 and beyond exits immediately, meaning that min_same and min_diff both remain at their initialized huge values.

When the loop is complete, the two minimum values are placed in their respective shared memory slots for the thread. Remember that each block has its own private shared memory that is visible to all threads in the block and invisible to everyone else. So at this point we have, for each block (row in our imagined matrix), the minimum for each thread (column in our imagined matrix).

In terms of our current *VarScreen* application, we have not accomplished anything very complex. Because we are assuming (though not demanding) that the number of cases does not exceed 16,384, all we've done is copy the distances from global memory to fast shared memory, although in doing so we have segregated the distances according to whether they are same class or different class. We do have to pause at a __syncthreads()

to make sure that all threads in the block have been processed, since CUDA does not guarantee any particular order in which warps will execute. This pause is not very wasteful, because such paused threads consume very little in the way of resources, and the scheduler will set other blocks to work.

Now comes the interesting part. Let's consider a single block (row in our imagined matrix). The next loop initializes j to 256>>1=128. The if(index<j) check means that only threads handling the first half of the columns in each row execute the next commands. We compare the element in this column j with the corresponding column exactly half of the row away, that at j+128, and we set the current column to whichever of the two is smaller. This is done separately for the same-class and different-class arrays. So when we have done a single pass through this loop, with each of 128 threads taking this action, we have reduced the needed number of columns in half. Now, in each row, the first 128 columns contain these pairwise minimums, and the second 128 columns can henceforth be ignored; we've gotten all that we need from them in this first pass through the loop.

Again, we pause until all threads in the block have completed and then move on to the second pass through the loop. Now j=64, and only the first 64 threads in each block take part. Each of the first 64 columns is compared to its partner 64 columns beyond and the smaller of the two placed in the current column. When we've finished this second pass through the loop, we have reduced the number of needed columns in half again, to 64.

This repeats until we reach the last pass, in which j=1. Now, in each block, only one thread takes part. We compare the first column with the second and place the smaller of the two in the first column. Voila, we have reduced the information down to just a single number in each row (block), and this number for each block is in that block's shared memory at [0].

The last step is to save the row-wise minimum to our allocated global memory areas for the same-class and different-class values. Recall that every thread in a block shares the same shared memory (good name, huh?), so any thread can be the one to do the store. It doesn't matter which one does it, but what matters very much is that only *one* of the threads does the store. Otherwise, we'll have contention over a store to global memory, not a good thing. Most people arbitrarily pick thread 0 to do the store.

Note that for clarity, this discussion has focused on a single row (metric) of the distance matrix, with that row determined by the *y* index of the block. But when we do the store, we have to pay attention to this detail and store our result in the correct location.

We're almost done, but not quite. If we had just one row in our imaginary matrix (at most REDUC_THREADS=256 cases), we would be done. But we probably have more cases than that, so we currently have a separate row minimum (referring to our imagined matrix) for each row/block. In order to finish the job, we need to pass through these blocks_per_grid rows and find their minimum. This is easily done with a second kernel:

```
__global__ void lfs_cuda_mindist_kernel_merge ( int blocks_to_merge )
{
  int i, metric ;
  float min_same, min_diff ;

  metric = blockIdx.x * blockDim.x + threadIdx.x ;
  if (metric >= d_ncases)
    return ;

  min_same = min_diff = 1.e30 ;
  for (i=0 ; i<blocks_to_merge ; i++) {
    if (d_minSame_out[i*d_ncols+metric] < min_same)
      min_same = d_minSame_out[i*d_ncols+metric] ;
    if (d_minDiff_out[i*d_ncols+metric] < min_diff)
      min_diff = d_minDiff_out[i*d_ncols+metric] ;
  }

  d_minSame_out[metric] = min_same ;
  d_minDiff_out[metric] = min_diff ;
}
```

The launch code for this second kernel is

```
warpsize = deviceProp.warpSize ;     // Threads per warp, likely 32 well into the future

threads_per_block = (ncases + warpsize - 1) / warpsize * warpsize ;
if (threads_per_block > 8 * warpsize)
  threads_per_block = 8 * warpsize ;

blocks_per_grid = (ncases + threads_per_block - 1) / threads_per_block ;

lfs_cuda_mindist_kernel_merge <<< blocks_per_grid , threads_per_block >>>
                              ( orig_bloc ks_per_grid ) ;
```

Remember that we have a metric (set of variable inclusion flags) for each case. We let this define the thread. This single dimension is all that we need for the launch. We pass as a parameter the number of blocks that we have to merge; we saved this from the launch of the first kernel.

This almost trivial kernel just passes through all of the original blocks, keeps track of the minimum, and then saves this result to the ultimate output area. So observe that these two arrays were allocated much larger than needed for their outputs so that they could also serve as scratch areas for the reduction algorithm of the first kernel.

Note that memory alignment is perfect in this kernel. Both work/output arrays were originally allocated on a cache-line boundary, each row is a multiple of 128 bytes (d_ncols), and adjacent threads access adjacent elements of these two vectors.

Computing the Terms for the Weight Equation

Equation (3.17) on Page 65 consists of an n_cases by n_cases matrix of terms, each of which is a negative exponential of a difference. We already have such a matrix of distances, and it's easy to convert this distance matrix into a matrix of terms. Here is the launch code, which has a thread for each case in the "case weight" dimension, and a block for each case in the metric dimension.

```
warpsize = deviceProp.warpSize ;    // Threads per warp, likely 32 well into the future

threads_per_block = (ncases + warpsize - 1) / warpsize * warpsize ;
if (threads_per_block > 8 * warpsize)
   threads_per_block = 8 * warpsize ;

block_launch.x = (ncases + threads_per_block - 1) / threads_per_block ;
block_launch.y = ncases ;
block_launch.z = 1 ;

lfs_cuda_term_kernel <<< block_launch , threads_per_block >>> ( iclass ) ;
```

The kernel is shown in the following and a discussion follows on the next page.

```
__global__ void lfs_cuda_term_kernel ( int iclass )
{
  int jcase ;
  float *dist_ptr, mindist ;
```

```
jcase = blockIdx.x * blockDim.x + threadIdx.x ;
if (jcase >= d_ncases)
   return ;

if (d_class[jcase] == iclass)
   mindist = d_minSame_out[blockIdx.y] ;
else
   mindist = d_minDiff_out[blockIdx.y] ;

dist_ptr = d_dist + blockIdx.y * d_ncols + jcase ; // distance[metric,jcase]
*dist_ptr = exp ( mindist - *dist_ptr ) ;
}
```

We pass as a parameter the class of the current case i. This is because Equation (3.17) has to subtract from each distance either the minimum same-class or the minimum different-class distance, according to whether the "other" class j is or is not in the same class as the current case i.

We choose as the appropriate minimum either the same-class minimum if cases i and j are in the same class or the different-class minimum otherwise. This minimum, of course, depends on the metric being done (k in Equation (3.17), the block's y coordinate in the code). Finally, we just point to the correct element in the distance matrix and convert that distance into a term in the weight computation matrix.

Transposing the Term Matrix

Memory alignment is perfect in the kernel just shown. Unfortunately, in order to compute the weights, we would need to sum that matrix down each column, across the rows. That operation would entail some most un-CUDA-like computation! So even though transposing a matrix in global memory can be relatively time consuming due to memory stalls, we really need to do it in order to be able to sum across columns using the extremely efficient reduction algorithm. It would be lovely to transpose it in place to save memory, but that operation is quite complex, and modern CUDA devices have tons of memory, so I chose to allocate a separate memory area. Here is that very simple kernel; I omit the trivial launch code.

```
__global__ void lfs_cuda_transpose_kernel ()
{
  int jcase ;
  float *term_ptr, *trans_ptr ;

  jcase = blockIdx.x * blockDim.x + threadIdx.x ;
  if (jcase >= d_ncases)
    return ;

  term_ptr = d_dist + blockIdx.y * d_ncols + jcase ;     // term[metric,jcase]
  trans_ptr = d_trans + jcase * d_ncols + blockIdx.y ;  // trans[jcase,metric]
  *trans_ptr = *term_ptr ;
}
```

Summing the Terms for the Weights

The final computation step is to sum each row of the transposed term matrix. We'll use the same reduction algorithm that we used for computing minimum distances. Here is the code for the reduction kernel. The principle is exactly the same as we saw in the minimum-distance routine, except that there we found the minimum of terms while here we sum terms. But note that we borrow d_minSame_out to compute and output the weights. We no longer need that array, so we might as well use it rather than allocate a separate weight array.

```
__global__ void lfs_cuda_sum_kernel ()
{
  __shared__ float partial_sum[REDUC_THREADS] ;
  int i, index ;
  float sum, *term_ptr ;

  index = threadIdx.x ;    // Associated with metric

  term_ptr = d_trans + blockIdx.y * d_ncols ; // This case

  sum = 0.0f ;
  for (i=blockIdx.x*blockDim.x+index ; i<d_ncases ; i+=blockDim.x*gridDim.x)
    sum += term_ptr[i] ;
```

```
  partial_sum[index] = sum ;
  __syncthreads() ;

  for (i=blockDim.x>>1 ; i ; i>>=1) {
    if (index < i)
      partial_sum[index] += partial_sum[index+i] ;
    __syncthreads() ;
    }

  if (index == 0) // We borrow d_minSameOut
    d_minSame_out[blockIdx.x*d_ncols+blockIdx.y] = partial_sum[index] ;
}
```

That was the first of the two kernels for summing, the reduction phase. Here is the kernel for summing the blocks that remain from the reduction, and the launch code follows.

```
__global__ void lfs_cuda_sum_kernel_merge ( int blocks_to_merge )
{
  int i, jcase ;
  float sum ;

  jcase = blockIdx.x * blockDim.x + threadIdx.x ;
  if (jcase >= d_ncases)
    return ;

  sum = 0.0 ;
  for (i=0 ; i<blocks_to_merge ; i++)
    sum += d_minSame_out[i*d_ncols+jcase] ;

  d_minSame_out[jcase] = sum / d_ncases ; // This is the final weight
}
  blocks_per_grid = (ncases + REDUC_THREADS - 1) / REDUC_THREADS ;
  if (blocks_per_grid > REDUC_BLOCKS)
    blocks_per_grid = REDUC_BLOCKS ;
  orig_blocks_per_grid = blocks_per_grid ;

  block_launch.x = blocks_per_grid ;   // Metric
  block_launch.y = ncases ;            // Case
  block_launch.z = 1 ;
```

```
lfs_cuda_sum_kernel <<< block_launch , REDUC_THREADS >>> () ;
cudaDeviceSynchronize() ;

warpsize = deviceProp.warpSize ;    // Threads per warp, likely 32 well into the future

threads_per_block = (ncases + warpsize - 1) / warpsize * warpsize ;
if (threads_per_block > 8 * warpsize)
  threads_per_block = 8 * warpsize ;

blocks_per_grid = (ncases + threads_per_block - 1) / threads_per_block ;


lfs_cuda_sum_kernel_merge <<< blocks_per_grid , threads_per_block >>>
                              ( orig_bloc ks_per_grid ) ;
```

Moving the Weights to the Host

All of the computation is done. All that remains is to move the weights, which are float, to the host, where they are double. This code is trivial. We've computed only n_cases weights, but we rounded that up to ncols. Of course we need to fetch only n_cases weights, but in deference to some possible future implementation that may demand that we fetch as many as were allocated (unlikely but very cheap insurance), I actually fetch all ncols of them. Feel free to call this silly and change it. It probably is. Then we just copy the weights, thereby converting them from float to double. It is highly unlikely that this copy operation will fail, but responsible programmers check for failure and handle it gracefully.

```
int lfs_cuda_get_weights ( double *weights , char *error_msg )
{
  int i, memsize ;
  char msg[256] ;
  cudaError_t error_id ;

  memsize = ncols * sizeof(float) ;

  error_id = cudaMemcpy ( weights_fdata , h_minSame_out , memsize ,
                            cudaMemcpyDeviceToHost ) ;
```

```
  for (i=0 ; i<ncases ; i++)
    weights[i] = weights_fdata[i] ;

  if (error_id != cudaSuccess) {
    // Handle this error
    }

return 0 ;
}
```

An Example of Local Feature Selection

I created a dataset consisting of about 4000 cases and 10 variables, X0 through X9. Each random variable is uniformly distributed on [–1, 1]. Variables X3 and X4 determine the class. A case is in one class if X3 and X4 are both positive or both nonpositive. The case is in the other class if one of these variables is positive and the other is not. This is a very difficult problem for many feature selection algorithms because the marginal distributions of these variables are identical for both classes, and the nature of the relationship between one of the variables with the class is determined by the value of the other variable. Here is the output of the LFS algorithm:

```
*****************************************
*                                       *
* Computing Local Feature Selection for predictor subset *
*     10  predictor candidates          *
*      5  predictors at most will define a metric space *
*      2  target bins                   *
*      3  iterations of LFS algorithm   *
*    500  random trials for real-to-binary f conversion *
*     20  trial values for beta optimization *
*    100  replications of complete Permutation Test *
*                                       *
*****************************************
```

-----------------> Percent of times selected <-----------------

Variable	Pct	Solo pval	Unbiased pval
X3	96.26	0.0100	0.0100
X4	69.62	0.0100	0.0100
X0	4.66	1.0000	1.0000
X1	2.94	1.0000	1.0000
X6	2.29	1.0000	1.0000
X7	1.76	1.0000	1.0000
X9	1.13	1.0000	1.0000
X8	0.58	1.0000	1.0000
X2	0.53	1.0000	1.0000
X5	0.39	1.0000	1.0000

It's a little curious that X3 was selected somewhat more often than X4, when they have identical roles in predicting the class, but I've seen this happen often. It's undoubtedly a random occurrence that would change with a different random set of cases. What is certainly clear is that these two variables are selected vastly more often than their worthless competitors. Also, the computed solo and unbiased p-values are impressive, leaving no doubt about the conclusion reached by the algorithm.

A Note on Runtime

This Local Feature Selection algorithm does have one downside that can make it unusable in some situations. Its runtime is proportional to the *cube* of the number of cases. On modern computers, especially those containing CUDA-capable video hardware, handling several thousand cases should be manageable. But if you get up to the range of many thousand cases, runtime will become so slow as to be impractical.

CHAPTER 4

Memory in Time Series Features

This chapter presents an approach to feature selection that is quite different from most other selection methods. In most applications, measured values of predictor variables (features) are directly associated with measured values of target variables (which may be numeric or class membership). Traditional feature selection looks for demonstrable relationships between predictor candidates and one or more targets, treating each case (set of measured values) as an independent sample.

In this chapter, we look at a powerful feature selection algorithm in which we take advantage of situations in which the samples are not independent. In particular, we assume the presence of *memory* in a time series of feature samples. This memory takes the form of a *hidden Markov model*. The fine mathematical details of hidden Markov models are beyond the scope of this text and widely available elsewhere. In this chapter, we will focus on the essential elements of this model, especially in the context of feature selection.

Traditional feature selection algorithms assume a direct relationship between predictors and a target. But in this chapter, we posit an underlying condition, the *state* of the process under study, which impacts both the predictors and the target. This process is assumed to exist at all times in exactly one of two or more possible states. The state at any given time impacts the distribution of associated variables. Some of these variables may be observable at the present time (predictors), while others may be unknown at the present time but be of great interest (targets). Our goal is to use measured values of the observable variables to determine (or make an educated guess at) the state of the process and then use this knowledge to estimate the value of an unobservable variable (the target) which interests us.

A hidden Markov model assumes a sequential process with an important property: the probability of being in a given state at an observed time depends on the process's state at the prior observed time. This is the nature of the memory inherent in a hidden Markov model.

This memory is immensely useful in some applications. For example, it may prevent whipsaws in a decision-making application. Suppose a certain state tends to be persistent in real life. Ordinary prediction methods will suffer if there is large random noise in the observed variables, which may snap the decision back and forth at the whim of chance. But the memory inherent in a hidden Markov model will tend to hold its decision in a persistent state even as noise in the measured variables tries to whip the decision back and forth. Of course, the downside of this memory is a tendency toward delayed decisions; the model may need several observed values to confirm a state change. But this is often a price well worth paying, especially in high-noise situations.

One application of a hidden Markov model is the prediction of a financial market. Perhaps the developer assumes that it is always in either a bull market (a long-term uptrend), a bear market (a long-term downtrend), or a flat market (no long-term trend). By definition, bull and bear markets cover an extended time period; one does not go from a bull to a bear market in one day and then return to a bull market the next day. Such direction changes are just short-term fluctuations in a more extensive move. If one were to use frequent observations to make daily predictions of whether the market is in a bull or bear state, these decisions could reverse ridiculously often. One is better off taking advantage of the memory of a hidden Markov model to stabilize behavior.

A Gentle Mathematical Overview

This section provides a rough overview of the mathematics involved. I have deliberately sacrificed strict rigor in order to favor intuition, making this material accessible to people who have only a modest mathematical background. For example, quantities that I call *likelihoods* are really not quite that due to normalization required for computational stability. But they are very similar, function the same way, and provide correct results. Also, I do not distinguish between joint and conditional probabilities. My only purpose here is to provide the equations that are key to the algorithm, along with intuitive justification. We begin with my nomenclature.

- We have T observations $x_0, x_1, x_2, ...x_{T-1}$. Throughout this section we will assume that each observation is a real number. However, it should always be clear that they could just as well be real-valued vectors, or class membership codes.

- The underlying process will always be in exactly one of N possible states. In a financial market prediction application, we may choose to assume three states: *Bull*, *Bear*, and *Flat*. In practice, we should assume as few states as possible, because runtime and computational fragility both blow up rapidly as the number of assumed states increases.

- The state of the process at time t is defined as $q(t)$, where t ranges from 0 to $T-1$ and $q(t)$ can take the values 1, 2, ..., N.

- The probability of transitioning from state i to state j is defined as a_{ij}. When $i=j$ we have the probability that the process will remain in the same state. This definition is expressed as shown in Equation (4.1).

$$a_{ij} = P\big(q(t) = j \mid q(t-1) = i\big) \tag{4.1}$$

- When the process is in state i, the observations x will follow the distribution whose probability density function is $f_i(x)$. Each state has its own distribution for the observations. If the observations take discrete values, these density functions will be actual probabilities, while if the observations are real numbers or real-valued vectors, they will be density functions rather than probabilities.

- The first observation (and only the first) x_0 has a set of prior probabilities for being in each possible state. These are written as $p_0(1), p_0(2), ..., p_0(N)$. In a financial market prediction application, we would likely set these prior probabilities according to our knowledge of historical market behavior around time $t=0$. If we have no basis for setting prior probabilities, then Bayes' *Principle of Equidistribution of Ignorance* says that we should make them equal.

A hidden Markov model is completely defined by three properties:

1)   The N initial probabilities $p_0(1)$, $p_0(2)$, ..., $p_0(N)$

2)   The set of N^2 transition probabilities a_{ij} for i, j from 1 through N

3)   The set of probability density functions f_i for i from 1 through N

The Forward Algorithm

Suppose we are at time t and we have observations up to and including this time: $x_0, x_1, ..., x_t$. Also suppose that we know all parameters of the model, the three properties just listed. Then, in the case of discrete observations, we can compute the probability of having obtained the observed values. And in the continuous case (the observations are real numbers or real vectors), we can compute the *likelihood* of having obtained that set of observations. This section will show a stable way of computing an effective proxy for that likelihood (which we will still call likelihood, somewhat incorrectly; see Page 111 for the strictly correct version).

But first, consider why we might want to compute this likelihood (or probability). After all, we have the set of observations in hand. Why are we interested in how likely that set is? The answer is that our ultimate goal is to reverse the order of model properties giving likelihoods; instead, we want to use observation likelihood to imply model properties. This well-known statistical technique is called *maximum likelihood* estimation. In the first paragraph of this section, we had this statement: "Also suppose that we know all parameters of the model, the three properties just listed." But we don't know the parameters! All we know are the observations, and from these we want to estimate the parameters of the model.

The principle behind maximum likelihood estimation is simple: we find the set of model parameters that maximizes the likelihood of the observations. Roughly stated, this means that we find the single model that, of all possible models, makes our obtained observations most likely.

Back to the subject at hand. We are at time t and we possess the observations through this time. We define $\alpha_t(i)$ to be the likelihood of having obtained these observations under the condition that at this time t we are in state i. This is shown in Equation (4.2). Naturally, we don't know what state we are in; it's not called a *hidden* Markov model for nothing! Also note that this quantity is the Greek letter alpha "α"

as opposed to the letter "a," which represents the transition probabilities as already shown. The letter "L" is used in this equation to emphasize that this is a likelihood, not a probability, in the continuous case.

$$\alpha_t(i) = L\big(x_0, x_1, \ldots, x_t \big| q(t) = i\big) \tag{4.2}$$

We have already defined the initial observation's probability of being in state i to be $p_0(i)$. This can be generalized to the probability of being in state i when we are at time t, *given the observations through that time*. Bayes' theorem gives us Equation (4.3). From here on, to save notational excesses, we will refer to just $p_t(i)$, omitting the joint behavior with prior observations but understanding its presence.

$$p_t\big(i \big| x_0, \ldots x_t\big) = \frac{\alpha_t(i)}{\sum\limits_{j=1}^{N} \alpha_t(j)} \tag{4.3}$$

Recall the basic conditional probability rule $P(A) = P(A|B) \bullet P(B)$, which also applies to likelihoods here. This lets us compute $\alpha_0(i)$, the likelihood associated with the first observation, conditional on the process being in state i, as shown in Equation (4.4). In this equation, $f_i(x_0)$ is the probability density of x_0 conditional on being in state i, and $p_0(i)$ is the probability of being in state i.

$$\alpha_0(i) = f_i(x_0) p_0(i) \tag{4.4}$$

This initial observation may have arisen from the process being in any of the N possible states, so the likelihood associated with the first observation, x_0, is the sum of the individual likelihoods, as shown in Equation (4.5).

$$L(x_0) = \sum_{i=1}^{N} \alpha_0(i) \tag{4.5}$$

We're on our way to finding the likelihood of observations through time t; we have it for the first observation. Hey, it's a start. On to the second. Actually, we'll be more general than that. We'll use recursion to get $\alpha_{t+1}(i)$ from $\alpha_t(i)$.

We can use Equation (4.3) to compute the probability of being in each state as of time t, always keeping in mind that this is not an absolute probability; it is in conjunction with the prior observations. This equation will play a key role in the recursion.

When we computed the likelihood of the first observation, we were lucky in that the model itself specified initial probabilities $p_0(1)$, $p_0(2)$, ..., $p_0(N)$. But after the first observation, we're on our own. We need an analog to Equation (4.4) but for time $t+1$ instead of time 0. The problem is that we may have arrived at state i from any of the N possible states at the prior time, t. So we have to sum the likelihoods due to coming from each of these possible states.

The probability of landing in state i at time $t+1$ from state j at time t is the probability of being in state j at time t times the probability of transitioning from state j to state i. The former is $p_t(j)$, which we have, and the latter is a model parameter. We must sum these across all N possible values of j. Thus, our recursive analog of Equation (4.4) is given by Equation (4.6).

$$\alpha_{t+1}(i) = f_i(x_{t+1}) \sum_{j=1}^{N} \left[p_t(j) a_{ji} \right] \tag{4.6}$$

Exactly as we did for the first observation in Equation (4.5), we can write the likelihood for this observation x_{t+1}, given the prior observations, as shown in Equation (4.7).

$$L(x_{t+1}) = \sum_{i=1}^{N} \alpha_{t+1}(i) \tag{4.7}$$

By the product rule, the net likelihood for the complete set of observations from the first through the last is the product of the individual likelihoods given by Equation (4.7). However, we don't dare compute it this way. The reason is that these likelihoods can be very small, especially if the data is not explained well by a hidden Markov model. The product of these terms will blow down to a machine zero in the blink of an eye. Therefore, to compute the net likelihood, we take the log of Equation (4.7) for each case and sum the logs. This is standard practice in maximum likelihood situations and is nearly always mandatory for anything other than tiny toy problems.

It's time to present code for the forward recursion algorithm. This code is in HMM.CPP. Details concerning the use of the HMM class will appear later. For now we deal with only the forward recursion algorithm. The forward() routine is called with a transition matrix these_transitions. Other necessary items have already been computed and are available as private members of the class. Here are the first few lines; a discussion follows.

```
double HMM::forward ( double *these_transitions )
{
  int i, j, t ;
  double sum, denom, log_likelihood, *prior_ptr, *trans_ptr ;

  denom = 0.0 ;
  for (i=0 ; i<nstates ; i++) {
    alpha[i] = init_probs[i] * densities[i*ncases] ;   // Equation (4.4)
    denom += alpha[i] ;                                // Equation (4.5)
    }

  log_likelihood = log(denom) ;
  for (i=0 ; i<nstates ; i++)
    alpha[i] /= denom ;                                // Equation (4.3)
```

In the preceding code, densities is an nstates (N) by ncases (T) matrix that already contains $f_i(x_t)$ for all N states (rows in the matrix) and cases (columns in the matrix). Thus, densities[i*ncases] is $f_i(x_0)$, the probability density function for the first case, conditional on being in state i. The array init_probs contains the initial case state probabilities, $p_0(i)$, part of the complete parameter set for the model. When we use Equation (4.3) to convert the likelihoods to probabilities, we store the result right back in alpha; there is no need to keep a separate probability array.

The recursion now begins. The most important point in this code is that the line that is commented out, which shows what the next line is doing, is inside loops nested three deep. And we'll see later that forward() is itself called inside a huge loop. This multiply/ sum operation is the largest eater of CPU time in the entire maximum likelihood optimization algorithm, so it should be made as fast and efficient as possible. I believe that the way I've programmed it with pointers is as good as can be done, but feel free to try alternatives.

```
for (t=1 ; t<ncases ; t++) {
  prior_ptr = alpha + (t-1) * nstates ;
  denom = 0.0 ;
  for (i=0 ; i<nstates ; i++) {
    trans_ptr = these_transitions + i ;
    sum = 0.0 ;
```

```
      for (j=0 ; j<nstates ; j++) {
//         sum += alpha[(t-1)*nstates+j] * these_transitions[j*nstates+i] ;
         sum += prior_ptr[j] * *trans_ptr ; // Summation inside Equation (4.6)
         trans_ptr += nstates ;
         }
      alpha[t*nstates+i] = sum * densities[i*ncases+t] ;   // Equation (4.6)
      denom += alpha[t*nstates+i] ;                        // Equation (4.5)
      }

   log_likelihood += log(denom) ;
   for (i=0 ; i<nstates ; i++)
      alpha[t*nstates+i] /= denom ;                        // Equation (4.3)
   }

   return log_likelihood ;
}
```

The Backward Algorithm

Just as we can move forward through time, recursing from the first observation, we can also move backward through time, recursing from the last observation. We begin by defining the reverse-computed (proxy) likelihoods as shown in Equation (4.8). The likelihood at time t encompasses cases from the next observation through the last.

$$\beta_t(i) = L\left(x_{t+1}, x_{t+2}, \ldots, x_{T-1} \middle| q(t) = i\right) \tag{4.8}$$

We have to make the apparently arbitrary but necessary starting likelihood at time $T-1$ (the final observation) to be 1 for all states: $\beta_{T-1}(i) = 1$ for all i. From there we recurse backward as shown in Equation (4.9). The p terms are obtained by normalizing the β terms at each step, exactly as was done with Equation (4.3) in the forward algorithm. This is true even in the first step, so $p_{T-1}(i) = 1/N$ for all i.

$$\beta_t(i) = \sum_{j=1}^{N}\left[f_j(x_{t+1})p_{t+1}(j)a_{ij}\right] \tag{4.9}$$

This is not the forum for a rigorous derivation of that equation, but the intuition is reasonable. At time t, if we are in state i, what is the likelihood of the subsequent set of observations? We can transition to any of the N possible states, so we have to compute a weighted sum of the likelihoods for each subsequent state. That likelihood is the probability density function of the next observation under the condition of being in that state, times the probability of being in that state (conditioned on all subsequent observations). Those are the first two terms in the product. Then we have to weight that contribution by the probability of transitioning from the current state i to that next state j. That's the third term.

There's one more item of interest, though of no importance to our algorithm. We can terminate the recursion to complete the net likelihood for the entire dataset, as shown in Equation (4.10).

$$L_{Termination} = \sum_{j=1}^{N}\left[p_0(j)f_j(x_0)\beta_0(j) \right] \qquad (4.10)$$

Here is the code for the backward algorithm. We begin by (explicitly for clarity, despite perhaps being overkill) initializing the last case's beta to 1 for all states, computing its log likelihood, and rescaling beta to probabilities, just as we did in the forward algorithm.

```
double HMM::backward ()
{
  int i, j, t ;
  double sum, denom, log_likelihood ;

  denom = 0.0 ;
  for (i=0 ; i<nstates ; i++) {
    beta[(ncases-1)*nstates+i] = 1.0 ;
    denom += beta[(ncases-1)*nstates+i] ;
    }

  log_likelihood = log(denom) ;
  for (i=0 ; i<nstates ; i++)
    beta[(ncases-1)*nstates+i] /= denom ;
```

We now recurse, using Equation (4.9). There is no need to take particular care about speed in this algorithm, as it is called a tiny fraction of the times that forward() is called.

```
for (t=ncases-2 ; t>=0 ; t--) {
  denom = 0.0 ;
  for (i=0 ; i<nstates ; i++) {
    sum = 0.0 ;
    for (j=0 ; j<nstates ; j++)
      sum += transition[i*nstates+j] * densities[j*ncases+t+1] * beta[(t+1)*nstates+j] ;
    beta[t*nstates+i] = sum ;
    denom += beta[t*nstates+i] ;
    }
  log_likelihood += log(denom) ;
  for (i=0 ; i<nstates ; i++)
    beta[t*nstates+i] /= denom ;
  }
```

As far as our application goes, this is all that we need; we could stop here. However, it is cheap to take the final step of terminating the recursion with Equation (4.10) to get the log likelihood of the complete dataset.

```
sum = 0.0 ;
for (i=0 ; i<nstates ; i++)
  sum += init_probs[i] * densities[i*ncases] * beta[i] ;

return log(sum) + log_likelihood ; // Include the termination likelihood
}
```

There is only one reason for terminating the recursion to get the complete log likelihood, but it's a good one. *The log likelihood computed this way should equal that computed with the forward algorithm.* (Of course, right? They are both the log likelihood of the complete dataset, so they should be the same!) Responsible programmers will compare these two numbers internally. If they differ by anything more than trivial amounts attributable to floating-point errors, then either the program has an overt error, or numerical stability issues are taking an undue toll and must be addressed.

Correct Alpha and Beta, for Those Who Care

As I hinted at several times, I told a little lie in the prior two sections. I called the computed alpha and beta values likelihoods, when really they are not exactly that. The problem is that by placing the probability-rescaled values (Equation (4.3) on Page 105) right back in the alpha and beta arrays, we destroy their property of being likelihoods. True likelihoods are not rescaled this way.

On the other hand, this is an easily forgiven lie. The reason is that, for our purposes, they behave in *exactly* the same way as true likelihoods. When subsequently used in clean floating-point operations, they provide exactly the same results as would be provided by true likelihoods. And most important of all (and my reason for this minor subterfuge) is that under not-so-clean floating-point operations, my rescaled alpha and beta are much, much more accurate. In fact, that's an understatement. If your application has thousands of cases, computing "correct" alphas and betas will produce such massive numerical instability that it will be virtually impossible to obtain correct results without special scaling later. I'll present code for "correct" computation, briefly explain it, and then show why "correct" computation is potentially dangerous in real life.

The key to "correct" computation is to keep a separate work array where the probability normalization is performed, but refrain from returning these normalized values to the alpha and beta matrices. We begin by initializing the first case exactly as we did in the earlier algorithm.

```
double HMM::forward ( double *these_transitions ) // Used by both initialize() and
estimate()
{
  int i, j, t ;
  double sum, denom, log_likelihood, *trans_ptr, work[MAX_STATES],
       temp[MAX_STATES] ;

  denom = 0.0 ;
  for (i=0 ; i<nstates ; i++) {
    work[i] = init_probs[i] * densities[i*ncases] ;
    denom += work[i] ;
  }
```

We compute the log likelihood of the first case as we did before, and we also normalize this alpha almost as we did before, though with a few differences:

- We normalize the version in the work array but don't put it in alpha.

- As will become obvious soon, alpha itself could quickly blow down to a machine zero. Thus, we actually store the log of alpha.

- Because as iterations (discussed later) proceed, init_probs as well as densities (and their product!) can become machine zero, we dare not just blithely take logs. We have to first verify that the quantity whose log we are taking is not zero and set its log to a very negative number as an alternative.

- Because we rescaled work before putting its log into alpha, we have to compensate for this rescaling by adding the log likelihood, which is equivalent to multiplying the true alpha by the scale factor denom.

```
log_likelihood = log(denom) ;
for (i=0 ; i<nstates ; i++) {
  work[i] /= denom ; // Equation (4.3) on Page 105
  if (work[i] > exp(-100.0))
    alpha[i] = log(work[i]) + log_likelihood ;
  else
    alpha[i] = -100.0 + log_likelihood ;
}
```

We now recurse forward over the remaining cases. If you compare this code to the prior, intelligently scaled code, you will see that the log likelihood is computed here with exactly the same algorithm as was done in the prior code, and so you will get identical results. In this code we keep in work exactly the same quantities that we would have placed in alpha.

One thing to note here is that at the start of each case, we must copy the recursed work array to a temp array. This is because we will need it in the internal summation, so we must not tamper with it while also using its values in the summation.

```
for (t=1 ; t<ncases ; t++) {
  denom = 0.0 ;
  for (j=0 ; j<nstates ; j++)
    temp[j] = work[j] ;
```

Here is the code for using Equation (4.6) on Page 106 to recurse from time t–1 to time t. If you compare this code to the prior code, keeping in mind the four points made a moment ago in regard to initializing for the first case, you should have no trouble understanding this.

```
for (i=0 ; i<nstates ; i++) {
  trans_ptr = these_transitions + i ;
  sum = 0.0 ;
  for (j=0 ; j<nstates ; j++) {   // Inner sum of Equation (4.6) on Page 106
    sum += temp[j] * *trans_ptr ;
    trans_ptr += nstates ;
    }
  work[i] = sum * densities[i*ncases+t] ;   // Equation (4.6)
  denom += work[i] ;
  }
log_likelihood += log(denom) ;
for (i=0 ; i<nstates ; i++) {
  work[i] /= denom ;                  // Equation (4.3) on Page 105
  if (work[i] > exp(-100.0))         // Don't take log of zero!
    alpha[t*nstates+i] = log(work[i]) + log_likelihood ;
  else
    alpha[t*nstates+i] = -100.0 + log_likelihood ;
  }
 }

return log_likelihood ;
}
```

I'll present the backward algorithm with little commentary, because its changes from the prior version precisely mirror the changes just seen for the forward algorithm.

```
double HMM::backward ()
{
  int i, j, t ;
  double sum, denom, log_likelihood, work[MAX_STATES], temp[MAX_STATES] ;
```

```
/*
   Initialize for last case (t=ncases-1):
   beta[ncases-1,i] = 1.0
*/

   denom = 0.0 ;
   for (i=0 ; i<nstates ; i++) {
     work[i] = 1.0 ;
     denom += work[i] ;
     }

   log_likelihood = log(denom) ;
   for (i=0 ; i<nstates ; i++) {
     work[i] /= denom ;
     beta[(ncases-1)*nstates+i] = 0.0 ; // log(1) = 0
     }
/*
   Recursion for remaining cases
*/

   for (t=ncases-2 ; t>=0 ; t--) {
     denom = 0.0 ;
     for (j=0 ; j<nstates ; j++)     // Copy so we don't mess with terms in summation
       temp[j] = work[j] ;
     for (i=0 ; i<nstates ; i++) {
       sum = 0.0 ;
       for (j=0 ; j<nstates ; j++)    // Equation (4.9) on Page 108
         sum += transition[i*nstates+j] * densities[j*ncases+t+1] * temp[j] ;
       work[i] = sum ;
       denom += work[i] ;
       if (sum > exp(-100.0))       // Don't take log of zero
         beta[t*nstates+i] = log(sum) + log_likelihood ;
       else
         beta[t*nstates+i] = -100.0 + log_likelihood ;
       }
```

```
     for (i=0 ; i<nstates ; i++)
        work[i] /= denom ;
     log_likelihood += log(denom) ;
     }
//  Termination:
   sum = 0.0 ;
   for (i=0 ; i<nstates ; i++)
     sum += init_probs[i] * densities[i*ncases] * work[i] ;

   return log(sum) + log_likelihood ;
}
```

There are two issues that I promised to address, and now is the time:

1)   In the "correct" alpha and beta algorithms, these quantities can
     quickly blow down to a machine zero, so rather than storing their
     actual values, we must store the log of their values.

2)   In real-life applications, the "correct" alpha and beta algorithms
     can be so numerically unstable as to render them dangerous
     unless special scaling precautions are taken when they are put to
     use later.

To address the first point, look at Equation (4.6) on Page 106, the forward recursion.
Every time we take a step, we are multiplying by a probability density. The same thing
happens in Equation (4.9) on Page 108, the backward recursion. These densities can and
very often will be quite small, maybe even minuscule. They keep getting cumulated into
the likelihood. In the "correct" code, we just add in the log likelihoods. But if we were
keeping the actual alpha and beta, rather than their logs, we would have to *multiply*
alpha and beta by these tiny values for each step. With even a hundred cases, we can
easily reach a machine zero, and with a thousand cases, this is virtually guaranteed. As a
result, we lose *all* accuracy in alpha and beta very soon. This is clearly unacceptable, so
we must cumulate the log of alpha and beta instead of the values themselves.

Now let's consider the second point. Suppose we do keep the logs of alpha and beta
(as we must!). This preserves their full accuracy in all but the most extreme pathological
situations. But that's often not good enough. There are various operations that we will
need to do later, such as computing state probabilities and updating the transition
matrix in an iterative process. In order to compute the individual terms needed for

115

these operations, we will need actual alpha and beta values. To get them we will need to exponentiate their logs, which is what we computed and stored. But if there are many cases, this exponentiation will evaluate to a machine zero. It doesn't matter if -3175.6 is the correct log of a term; $\exp(-3175.6)$ is still going to be zero as far as the computer is concerned. Later, we'll see a rescaling technique that largely solves that problem, but it's still a nuisance, and it's easy to neglect to do it right if we are not careful. All in all, we are much better off with the "incorrect" scaling, which is equivalent but vastly more stable with numerical computation.

Some Mundane Computations

Before continuing with the interesting stuff, we need to get some boring things out of the way. All of these routines are in HMM.CPP.

Means and Covariances

We all know how to compute means and covariances, so it's almost pointless to waste paper by presenting the code here. But it does illustrate the way these quantities are stored in the HMM object. We compute the means and covariances for the first state and then copy them to the other states. Here we compute the mean:

```
void HMM::find_mean_covar ()
{
   int i, j, istate, icase ;
   double *mean_ptr, *covar_ptr, *case_ptr, diff, diff2 ;

   for (istate=0 ; istate<nstates ; istate++) {
      mean_ptr = means + istate * nvars ;            // Mean vector for this state
      covar_ptr = covars + istate * nvars * nvars ;  // Symmetric covariance

      if (istate == 0) { // Compute means and covariances once (istate==0) then copy
         for (i=0 ; i<nvars ; i++) {
            mean_ptr[i] = 0.0 ;
            for (j=0 ; j<nvars ; j++)
               covar_ptr[i*nvars+j] = 0.0 ;
         }
```

```
    for (icase=0 ; icase<ncases ; icase++) {
      case_ptr = data + icase * nvars ;
      for (i=0 ; i<nvars ; i++)
        mean_ptr[i] += case_ptr[i] ;
      }

    for (i=0 ; i<nvars ; i++)
      mean_ptr[i] /= ncases ;
```

Now we continue on to the covariances. The covariance matrices are symmetric, so we need only compute one triangle and then copy it to the other triangle. But this routine is called only once and is very fast. My opinion is that it's more clear to compute the entire matrix despite the waste. Feel free to disagree and revise the code if you wish.

We keep one additional copy of the covariance matrix in init_covar. This will be used as the center of perturbation during the initialization search for good starting parameters.

```
    for (icase=0 ; icase<ncases ; icase++) {
      case_ptr = data + icase * nvars ;
      for (i=0 ; i<nvars ; i++) { // Symmetric, but do entire matrix for clarity; it's very fast
        diff = case_ptr[i] - mean_ptr[i] ;
        for (j=0 ; j<nvars ; j++) {
          diff2 = case_ptr[j] - mean_ptr[j] ;
          covar_ptr[i*nvars+j] += diff * diff2 ;
          }
        }
      }
    for (i=0 ; i<nvars ; i++) { // Symmetric, but do entire matrix for clarity; it's very fast
      for (j=0 ; j<nvars ; j++) {
        covar_ptr[i*nvars+j] /= ncases ;
        init_covar[i*nvars+j] = covar_ptr[i*nvars+j] ; // We'll use this for initialization
        }
      }
    } // If istate==0 (we must compute means and covariances)

  else { // After state 0, just copy covariances and means
    for (i=0 ; i<nvars ; i++) {
      mean_ptr[i] = means[i] ;
```

117

```
      for (j=0 ; j<nvars ; j++)
        covar_ptr[i*nvars+j] = covars[i*nvars+j] ;
      }
    } // If istate > 0
  } // For all states
}
```

Densities

Every time we change the means or covariances (during initialization as well as iterative improvement), we have to recompute the density matrix. This is an nstates by ncases matrix which contains the probability density function for every case under the condition of every possible state. This routine will be called many, many times and is the second-largest eater of computer time, second only to the forward recursion of forward(). So it behooves us to make it as efficient as possible. To be more specific, it is its call to mv_normal() that is the real beast. We'll tackle that routine in a moment. Here is the find_densities() code:

```
void HMM::find_densities ( double *these_means ) // Used by both initialize() and
estimate()
{
  int i, istate ;
  double *mean_ptr, *covar_ptr, *density_ptr, det ;

  for (istate=0 ; istate<nstates ; istate++) {        // Do each possible state
    mean_ptr = these_means + istate * nvars ;     // Mean vector for this state
    covar_ptr = covars + istate * nvars * nvars ;   // Symmetric covar for this state
    density_ptr = densities + istate * ncases ;      // Density vector for this state

    invert ( nvars , covar_ptr , inverse , &det , rwork , iwork ) ;   // This is very fast
    for (i=0 ; i<ncases ; i++)
      density_ptr[i] = mv_normal ( nvars, data+i*nvars, mean_ptr, inverse, det , rwork ) ;
    }
}
```

This code is trivial. For each state it inverts that state's covariance matrix to get what mv_normal() needs and then calls that routine once for each case. You don't need to be terribly particular about efficiency in the matrix inversion routine, because in the vast majority of applications there are very few variables (often just one or two), so inversion is fast and stable. The routine in INVERT.CPP returns 1 if the matrix is singular, which you may wish to check for complete thoroughness. On the other hand, in any well-designed application that I've ever encountered, singularity in a state's covariance would be rare to the point of being pathological, so I don't bother.

The Multivariate Normal Density Function

In my own work, I assume a multivariate normal distribution of the data. If you wish to use some other distribution, you will need to make appropriate changes. Be comforted that the forward and backward recursions remain exactly the same, so all you would need to change is this density function (easy), the initial parameter computations (medium) and the parameter update algorithms discussed later (possibly pretty tough).

The multivariate normal density function is defined in Equation (4.11). In this equation, μ is the mean vector, Σ is the covariance matrix, and there are k variables. The code is listed as follows, and comments follow on the next page.

$$f(\mathbf{x}) = \frac{\exp\left(-0.5*(\mathbf{x}-\mu)^{\mathbf{T}}\Sigma^{-1}(\mathbf{x}-\mu)\right)}{\sqrt{(2\pi)^{k}|\Sigma|}} \tag{4.11}$$

```
double mv_normal (
   int nv ,                  // Number of variables
   double *x ,               // Case to be evaluated (nv long)
   double *means ,           // Mean vector (nv long)
   double *inv_covar ,       // Inverse of covariance matrix, nv*nv, symmetric
   double det ,              // Determinant of covariance
   double *work )            // Work vector nv long
{
   int i, j ;
   double sum, temp ;
   double log_2pi = log ( 2.0 * 3.141592653589793 ) ;

   for (i=0 ; i<nv ; i++)
      work[i] = x[i] - means[i] ;
```

```
  sum = 0.0 ;

  // Do lower triangle
  for (i=1 ; i<nv ; i++) {
    for (j=0 ; j<i ; j++)
      sum += work[i] * work[j] * inv_covar[i*nv+j] ;
    }

  sum *= 2 ; // Include symmetric upper triangle

  // Do diagonal
  for (i=0 ; i<nv ; i++)
    sum += work[i] * work[i] * inv_covar[i*nv+i] ;

  if (sum > 500)
    sum = exp ( -0.5 * 500 ) ;
  else
    sum = exp ( -0.5 * sum ) ;
  temp = fabs(det) * exp ( nv * log_2pi ) ;   // Det can never be negative; see text below

  return sum / sqrt ( temp + 1.e-120 ) ;       // Temp should never be zero; see text below
}
```

The inverted covariance matrix is symmetric, so we need to do only one minor triangle, the lower triangle here. That triangle's contribution is doubled to include the neglected upper triangle. Then we do the diagonal.

It is conceivable that the sum could be so large that its negative exponential is a machine zero. Because a normal density can never be truly zero, we impose a tiny limitation on how small this exponential can be. This makes a zero result nearly impossible. Feel free to eliminate this test if you consider it unnecessary in your application.

I include two more paranoid extravagances. It is arguably ridiculous to take the absolute value of the determinant, because a true covariance matrix can never have a negative determinant. But I want to make sure we never trigger a math exception by taking the square root of a negative number that can (very rarely!) arise from numerical issues. Also, as long as the covariance matrix is nonsingular, the denominator will never be zero. But I add 1.e-120 as cheap and harmless insurance against triggering a math exception by dividing by zero. Feel free to eliminate either or both of these borderline crazy actions. Being a cautious guy, I like the comfort of including them.

Starting Parameters

The Baum–Welch algorithm for finding maximum-likelihood parameters, which is the method used in this text, enjoys a long history of widespread use. It is fast and reliable, common characteristics of expectation–maximization algorithms (which this is). However, there is one serious issue to be dealt with: although it is guaranteed to converge to the nearest local maximum, the likelihood surface is far from convex. The function is filled with suboptimal local maxima, so we dare not just pick any old random starting point and iterate from it. We *must* pick a starting point that has large likelihood compared to other points in the parameter space. Moreover, our trial parameter sets must be intelligently chosen.

This initialization is not a trivial undertaking. For a smallish problem (two variables, three states), 1000 random trials is an absolute minimum, with 10,000 not being unreasonable. Initialization typically takes ten or more times as much computer time as the iterative improvement that follows.

Outline of the Initialization Algorithm

The search for a good starting parameter set is performed in initialize() in the file HMM.CPP as shown in the following text. Repeat these steps many times:

1) Perturb the means, making sure that for each variable, the mean offset from the dataset mean across states is zero.

2) Perturb the common covariance matrix by shrinking the diagonal moderately and shrinking the off-diagonals even more.

3) Perturb the transition probabilities. Half of the time assign equal probabilities to all transitions. Otherwise, favor remaining in the same state.

4) Compute the densities, and then compute the log likelihood using the forward() algorithm.

5) Keep track of the best (maximum likelihood) parameter set.

Perturbing Means

For each variable, we will perturb its trial mean using the dataset mean as a center. Also, the amount we perturb will be based on the dataset standard deviation of the variable. This avoids the need for the user to provide standardized data; any center or variance of the data will be handled well.

    We also take care that the degree to which we perturb each variable sums to zero across all states. That way we make sure that the trials are not unbalanced, centering away from the dataset mean, which would usually make no sense.

```
for (i=0 ; i<nvars ; i++) {
  sum = 0.0 ;
  for (istate=0 ; istate<nstates ; istate++) {
    if (istate < nstates-1) {
      covar_ptr = covars + istate * nvars * nvars ;
      wt = 6.0 * (unifrand() - 0.5) * sqrt(covar_ptr[i*nvars+i]) ;
      sum += wt ;
      }
    else
      wt = -sum ;
    trial_means[istate*nvars+i] = means[istate*nvars+i] + wt ;
    }
  }
```

In this code, we take the square root of the covariance diagonal corresponding to the variable being handled. By subtracting 0.5 from a uniform (0,1) random number and multiplying by 6 to get the multiplier, we get an offset that is at most +/–3 standard deviations. We do this for all but the last state, cumulating the sum of the weights. For the last state, we use the negative of the sum from the prior states, which ensures that all weights sum to zero, thereby centering them across the states. This offset is added to the dataset mean to get the trial value for the model's mean for this variable and state.

Perturbing Covariances

Things are not quite so straightforward for generating trial covariance matrices. For trial means, we have the dataset means as a logical center of perturbation. But the dataset covariances are almost certainly inflated estimates of the individual state covariances.

This is because variation within the complete dataset includes not only variation within each individual state, which is what we want to estimate, but also variation across states due to differences in the state means. For example, suppose we have one variable and two states. Within each state, this variable may have a variance of 1. But suppose the mean in one state is 5 and the mean in the other state is 20. That variation within the complete dataset of values near 5 and values near 20 will cause the dataset variance to vastly exceed the within-state variance. In a multivariate case, correlations will inflate as well. So unlike the case for means, we cannot use the dataset covariance as a center of perturbation.

We do allow one simplification: we initialize the covariance matrix to be the same for all states. In many applications this is naturally part of the true model. If not, it is still almost always a reasonable starting point. To get a trial matrix, we shrink the diagonals by a random factor ranging from 0.4 to 0.9, and we shrink the off-diagonals by an additional random factor ranging from 0 to 0.7. Much practical experience indicates that starting with low implied correlation like this improves the starting points. It seems to be easier for the algorithm to add correlation than to take it away.

```
wtd = 0.4 + 0.5 * unifrand() ;   // Shrinkage for diagonal
wto = 0.7 * unifrand() * wtd ;   // Shrinkage for off-diagonal
for (i=0 ; i<nvars ; i++) {
  for (j=0 ; j<nvars ; j++) {
    if (i == j)
      dtemp = wtd * init_covar[i*nvars+j] ;
    else
      dtemp = wto * init_covar[i*nvars+j] ;
    for (istate=0 ; istate<nstates ; istate++) {
      covar_ptr = covars + istate * nvars * nvars ;
      covar_ptr[i*nvars+j] = dtemp ;
      }
    }
  }
```

Perturbing Transition Probabilities

We use two different methods for generating trial transition probabilities. Before commencing the trial loop, we initialize the trial values to be uniform, and we stick with these for the first half of all trials.

```
prob = 1.0 / nstates ; // For the first half of trials, assign equal transition probs
for (istate=0 ; istate<nstates ; istate++) {
  for (i=0 ; i<nstates ; i++)
    trial_transition[istate*nstates+i] = prob ;
  }
```

After having done half of the trials, we switch to a random method in which staying in the same state has a strong tendency to be favored. This "stick with it" behavior is common in many applications. In fact, for many applications, the constants of 0.4 and 0.5 are not strong enough. You may wish to adjust them to provide an even stronger tendency to stay in a state.

```
if (itrial >= n_trials/2) {
  for (istate=0 ; istate<nstates ; istate++) {
    prob = trial_transition[istate*nstates+istate] = 0.4 + unifrand() * 0.5 ;
    for (i=0 ; i<nstates ; i++) {
      if (i != istate)
        trial_transition[istate*nstates+i] = (1.0 - prob) / (nstates - 1.0) ;
      }
    }
  }
```

A Note on Random Number Generators

All of the code snippets just presented call unifrand() to provide uniform random deviates in (0,1). The quality of the random numbers is not as critical as it would be for volume-oriented applications such as Monte-Carlo integration; a well-crafted linear congruential generator should be fine. However, if this random initialization process is multithreaded, you *must* be sure to use a thread-safe generator! All kinds of bizarre errors can occur if you do not, including disastrous errors whose effects may be hidden. Chances are, you will not experience an obvious failure or complete crash. Rather, everything will seem to be working correctly, but you'll be generating poor trial parameter sets.

There are basically two methods for creating a thread-safe random generator. The easiest is to keep nothing in a static variable; just pass the seed as a parameter. Here is my workhorse lower-quality thread-safe generator, which must never be called with iparam=0 and which will never return exactly zero.

```
#define IA 16807
#define IM 2147483647
#define IQ 127773
#define IR 2836

double fast_unif ( int *iparam )
{
  long k ;

  k = *iparam / IQ ;
  *iparam = IA * (*iparam - k * IQ) - IR * k ;
  if (*iparam < 0)
    *iparam += IM ;
  return *iparam / (double) IM ;
}
```

The other approach, which I employ in my unifrand(), is to use the operating system's API to prevent thread contention. In Windows, this can be done by calling InitializeCriticalSection() before calling the generator and then using EnterCriticalSection() and LeaveCriticalSection() to guard the generator itself.

The Complete Optimization Algorithm

This section presents the remaining parts of the algorithm for finding maximum-likelihood parameters for a hidden Markov model. The overall flow of the algorithm is as follows:

1)   Compute the mean vector and covariance matrix of the complete dataset. This was described on Page 116.

2)   Initialize a starting parameter set by trying a large number of random values and choosing the best. This was described on Page 121.

3) While iterating until converged...

4) Compute densities for the current parameter set. This was described on Page 118.

5) Call forward() to compute the log likelihood and the alpha matrix. This was described on Page 104.

6) Call backward() to compute the beta matrix. This was described on Page 108.

7) For each observation, compute the probability that the process is in each of the possible states. This is described below.

8) For each state, update its mean vector and covariance matrix. This will be described on Page 130.

9) Update the initial state probability vector and the transition probability matrix. This will be described on Page 132.

10) Check for convergence. If not converged, loop back to step 4.

Computing State Probabilities

Before diving into this topic, I want to clarify where it falls in a more general view of the optimization algorithm. If the detailed algorithm shown on the prior page is distilled down to its essence, it comes down to just two steps that alternate until convergence is obtained. We begin with an educated guess for a good starting parameter set and then do the following:

1) Under the assumption that the current model parameters are correct, compute for each observation the probability that the process is in each of its possible states. (This is the topic of this section.)

2) Under the assumption that the set of observation-state probabilities is correct, update our estimate of the model parameters.

3) If we have not converged, loop back to step 1.

In other words, we just alternate back and forth: use the model parameters and dataset to compute state probabilities, and then use the state probabilities to compute updated model parameters. Repeat as needed. This algorithm is guaranteed to converge to a local optimum, which is hopefully the global optimum. This hope is the motivation for working hard to find a good starting estimate for the model parameters.

The algorithm for computing the state membership probability for each observation is almost trivial. Recall that, by definition, $\alpha_t(i)$ is the likelihood (my normalized version or the "correct" version) of state i associated with observations from the first through time t, and $\beta_t(i)$ is the likelihood of state i associated with observations from time $t+1$ through the last. Thus, their product is the likelihood of state i for the complete set of observations. Bayes' rule tells us that the probability that the observation at time t is in state i is given by Equation (4.12).

$$\gamma_t(i) = \frac{\alpha_t(i)\,\beta_t(i)}{\sum_{j=1}^{N}\left[\alpha_t(j)\,\beta_t(i)\right]} \tag{4.12}$$

If you are using my normalized version of alpha and beta, you can get away with simple code for computing this. The first loop sums the terms, computing the denominator of Equation (4.12). The second loop computes the probabilities.

```
for (icase=0 ; icase<ncases ; icase++) {

  sum = 0.0 ;
  for (istate=0 ; istate<nstates ; istate++) {
    temp = alpha[icase*nstates+istate] * beta[icase*nstates+istate] ;
    if (temp < 1.e-12)
      temp = 1.e-12 ;
    sum += temp ;
  }
  for (istate=0 ; istate<nstates ; istate++) {
    temp = alpha[icase*nstates+istate] * beta[icase*nstates+istate] ;
    if (temp < 1.e-12)
      temp = 1.e-12 ;
    state_probs[icase*nstates+istate] = temp / sum ;
  }
}
```

There are two reasons for limiting the product to 1.e-12. The important reason is that we do not want sum to have even the tiniest chance of being zero, because 0/0 will trigger a floating-point exception. Of course, due to the normalization of alpha and beta, I suspect that having the product be zero for *every* state is virtually impossible, so this is probably wasted effort. But by now you know that I'm a very cautious guy.

The second reason is arguable, and you may wish to shrink the limit to something more like 1.e-200. My own opinion is that we should never say that there is zero probability that the process will be in some state for some observation. By putting this floor under the product, we ensure that there is at least a minuscule probability for every state. Please feel free to disagree and lower the floor to barely above zero if you wish. Also note that soon, we will see a much more complex but effective way to eliminate the need for any probability floor at all (allowing zero probability), if that is your preference.

If you are computing the "correct" values of alpha and beta, the simple code just shown will fail disastrously. With even a hundred cases, there is a significant probability that *all* of those alpha–beta products will evaluate to a machine zero for at least some cases after being exponentiated to undo the fact that they are logs. By placing even the tiniest floor under the product, you will end up with a probability of $1/N$ for all states, and if you do not impose a floor, you will divide by zero. It's a lose–lose situation.

There is a way to rescale the products to avoid this problem. This method also takes care of the extremely remote possibility that this situation might arise with my scaled alpha and beta, and it essentially eliminates placing a floor under the probabilities. I do this in the HMM.CPP code. That code also shows how to do the scaling for "correct" alpha and beta. I won't discuss that here because I do not recommend the "correct" approach. But if that is your preference, just refer to the code in HMM.CPP.

You saw in the preceding code that the computation occurs in two phases. First we find the sum of the terms, and then we compute the probabilities by dividing the terms by their sum. Now we need to use three stages. The first stage computes a scaling factor that will eliminate numerical problems. The second and third stages are the same two stages that we just saw, except with rescaling applied. Here is the outermost case loop and the first stage:

```
for (icase=0 ; icase<ncases ; icase++) {
  for (istate=0 ; istate<nstates ; istate++) {   // Step 1 of 3: Find max of (alpha * beta)
    temp = alpha[icase*nstates+istate] * beta[icase*nstates+istate] ;
```

```
  if (temp < 1.e-100)
    temp = 1.e-100 ;
  if (istate == 0 || temp > max_prod)
    max_prod = temp ;
  }
```

In this code, we let temp be the numerator term of Equation (4.12). To avoid division by zero, we do place a floor under it, but as we will soon see, this does *not* place a floor under the probability; zero is still possible (which may or may not be a good thing, according to your opinion and the application). This initial loop finds the maximum value taken by any term for this observation.

The second step is to compute the denominator of Equation (4.12), which is just the sum of the terms. But the key here is that rather than summing the terms, we divide each term by the maximum, with the result that the maximum term in this sum is 1, regardless of how small the original term was.

```
sum = 0.0 ;
for (istate=0 ; istate<nstates ; istate++) {
  temp = alpha[icase*nstates+istate] * beta[icase*nstates+istate] ;
  if (temp < 1.e-100)
    temp = 1.e-100 ;
  sum += temp / max_prod ;
  }
```

Finally, we divide each term by their sum, completing Equation (4.12). But again, we first divide each term by the maximum term. This is why, even though we placed a floor under the individual terms, there is no effective floor under the probabilities computed here. If any terms are really minuscule, this division will result in a machine zero being generated as the quotient probability.

```
for (istate=0 ; istate<nstates ; istate++) {
  temp = alpha[icase*nstates+istate] * beta[icase*nstates+istate] ;
  if (temp < 1.e-100)
    temp = 1.e-100 ;
  state_probs[icase*nstates+istate] = (temp / max_prod) / sum ;
  }
}
```

Do note that in the extremely unlikely event that for some case all of the state terms are less than the 1.e-100 floor, we will get a probability of $1/N$ for every state. There is nothing unreasonable about this result, because such an event means only that this observation is wildly incompatible with the model, and so assigning equal probabilities to all states is as good as or even better than favoring some state that surely does not deserve favor.

Updating the Means and Covariances

After the state probability for each case has been computed, we must update the estimated mean and covariance matrix for each state. This is easier than one may think, because the algorithm is identical to the ordinary way means and covariance matrices are computed (e.g., what we saw on Page 116) except that for each observation and state, the contribution is weighted by the probability of being in that state. These probabilities are already in state_probs.

Each state is processed individually in the outermost loop. The first step for a state is to zero the mean vector and covariance matrix. Actually, we set the diagonal to a tiny value that will normally be washed out, just to prevent a zero diagonal (which would lead to disaster later). Then we compute the weighted mean.

```
for (istate=0 ; istate<nstates ; istate++) {
   mean_ptr = means + istate * nvars ;          // Mean vector for this state
   covar_ptr = covars + istate * nvars * nvars ; // Symmetric covariance matrix

   for (i=0 ; i<nvars ; i++) {
     mean_ptr[i] = 0.0 ;
     for (j=0 ; j<nvars ; j++) {
       if (i == j)
         covar_ptr[i*nvars+j] = 1.e-10 ;   // Prevent zero variance
       else
         covar_ptr[i*nvars+j] = 0.0 ;
     }
   }

   denom = 0.0 ;
   for (icase=0 ; icase<ncases ; icase++) {
     case_ptr = data + icase * nvars ;
```

```
    weight = state_probs[icase*nstates+istate] ;
    for (i=0 ; i<nvars ; i++)
       mean_ptr[i] += weight * case_ptr[i] ;
    denom += weight ;
    }

  for (i=0 ; i<nvars ; i++)
    mean_ptr[i] /= (denom + 1.e-100) ;
```

The covariance matrix is handled the same way. Note that the covariance matrix is symmetric, so we could compute just half and then copy that to the other half. But this operation is very fast and it's more clear to show it this way. Change it if you wish.

After computing the covariance matrix, we take one last step to head off problems from rare pathological situations. We pass through the off-diagonal elements and impose an upper limit on their magnitude of the harmonic mean of their corresponding diagonals. This ensures that the matrix is invertible, a property that we will need to compute densities.

```
  denom = 0.0 ;
  for (icase=0 ; icase<ncases ; icase++) {
    case_ptr = data + icase * nvars ;
    weight = state_probs[icase*nstates+istate] ;
    for (i=0 ; i<nvars ; i++) {          // Symmetric, but do entire matrix for clarity
       diff = case_ptr[i] - mean_ptr[i] ;
       for (j=0 ; j<nvars ; j++) {
         diff2 = case_ptr[j] - mean_ptr[j] ;
         covar_ptr[i*nvars+j] += weight * diff * diff2 ;
         }
       }
    denom += weight ;
    }
  for (i=0 ; i<nvars ; i++) {            // Symmetric, but do entire matrix for clarity
    for (j=0 ; j<nvars ; j++)
       covar_ptr[i*nvars+j] /= (denom + 1.e-100) ;
    }

  for (i=0 ; i<nvars ; i++) {            // Ensure matrix is invertable
    for (j=0 ; j<nvars ; j++) {
```

```
        if (i!=j && covar_ptr[i*nvars+j] > 0.999999 * sqrt ( covar_ptr[i*nvars+i] *
                                                        covar_ptr[j*nvars+j] ))
            covar_ptr[i*nvars+j] = 0.999999 * sqrt ( covar_ptr[i*nvars+i] *
                                                covar_ptr[j*nvars+j] );
        if (i!=j && covar_ptr[i*nvars+j] < -0.999999 * sqrt ( covar_ptr[i*nvars+i] *
                                                        covar_ptr[j*nvars+j] ))
            covar_ptr[i*nvars+j] = -0.999999 * sqrt ( covar_ptr[i*nvars+i] *
                                                covar_ptr[j*nvars+j] ) ;

        }
    }
} // For all states
```

Updating Initial and Transition Probabilities

The final model parameters to be updated are the initial probabilities $p_0(i)$ for $i=1,...,N$, and the transition probability matrix a_{ij} for $i,j=1,...,N$. The former is easy: it's just the state probabilities for the first case, which we already computed.

```
for (istate=0 ; istate<nstates ; istate++)
    init_probs[istate] = state_probs[istate] ;
```

The transition probability matrix is considerably more complicated. Consider a transition from state i to state j. We let $\zeta_t(i,j)$ designate the joint probability of being in state i at time t and state j at time $t+1$. We state without proof (widely available elsewhere) that this quantity is given by Equation (4.13). Then, to compute the updated transition probabilities, we average these across the dataset, normalizing to make sure that each row sums to one. This is given by Equation (4.14).

$$\zeta_t(i,j) = \frac{\alpha_t(i) a_{ij} f_j(x_{t+1}) \beta_{t+1}(j)}{\sum_{k=1}^{N}\sum_{l=1}^{N}\left[\alpha_t(l) a_{lk} f_k(x_{t+1}) \beta_{t+1}(k)\right]} \tag{4.13}$$

$$a_t(i,j) = \frac{\sum_{t=0}^{T-2}\zeta_t(i,j)}{\sum_{k=1}^{N}\left[\sum_{t=0}^{T-2}\zeta_t(i,k)\right]} \tag{4.14}$$

If you use my scaled alpha and beta, you can also use the relatively simple code shown below. I use it in much of my own software, and it is a straightforward implementation of the equations just shown. A walkthrough will appear on the following page. After that I'll present the much more complex scaled version that appears in HMM.CPP.

```
for (istate=0 ; istate<nstates ; istate++) { // Initialize updated transition matrix to 0
  for (jstate=0 ; jstate<nstates ; jstate++)
    trans_work2[istate*nstates+jstate] = 0.0 ;
  }

for (icase=1 ; icase<ncases ; icase++) {       // Sum terms inside Equation (4.14)
  sum = 0.0 ;                                    // Sum denominator of Equation (4.13)
  for (istate=0 ; istate<nstates ; istate++) {
    for (jstate=0 ; jstate<nstates ; jstate++) {   // Numerator terms of Equation (4.13)
      temp = alpha[(icase-1)*nstates+istate] *
        transition[istate*nstates+jstate] *
        densities[jstate*ncases+icase] *
        beta[icase*nstates+jstate] ;
      if (temp < 1.e-120)              // Prevent highly unlikely sum=0
        temp = 1.e-120 ;
      trans_work1[istate*nstates+jstate] = temp ;   // Save to avoid recomputing
      sum += temp ;                        // Cumulate denominator of Equation (4.13)
      }
    }
  for (istate=0 ; istate<nstates ; istate++) {     // Normalize Equation (4.13)
    for (jstate=0 ; jstate<nstates ; jstate++) {   // And sum Equation (4.14)
      temp = trans_work1[istate*nstates+jstate] ;
      trans_work2[istate*nstates+jstate] += temp / sum ;
      }
    }
  } // For icase, summing numerator terms of Equation (4.14)

for (istate=0 ; istate<nstates ; istate++) {
  sum = 0.0 ;                                      // Sums denominator of Equation (4.14)
  for (jstate=0 ; jstate<nstates ; jstate++) {
    temp = trans_work2[istate*nstates+jstate] ;    // Numerator term
```

```
    if (temp < 1.e-12)                    // Prevent highly unlikely zero denominator
      temp = 1.e-12 ;
    sum += temp ;                         // Cumulate denominator of Equation (4.14)
    }
  for (jstate=0 ; jstate<nstates ; jstate++) {
    temp = trans_work2[istate*nstates+jstate] ;
    if (temp < 1.e-12)                    // Duplicate action in loop above
      temp = 1.e-12 ;
    transition[istate*nstates+jstate] = temp / sum ; // Complete Equation (4.14)
    }
  }
```

The first step in the preceding code is to zero out the *a* matrix in preparation for summing each numerator term in Equation (4.14). The loop over icase then sums these individual terms.

Within this icase loop, we perform two steps. The first step computes the individual numerator terms of Equation (4.13). They are saved in trans_work1 to avoid the need to recompute them, and their sum is cumulated for the denominator of this equation.

The second step in the icase loop fetches the individual terms from trans_work1, divides them by their sum to complete Equation (4.13), and cumulates them in trans_work2 to build up the numerator terms of Equation (4.14).

After the icase loop is finished, we loop down the rows of the transition probability matrix. Inside this row loop, we perform two tasks that should be familiar by now. First we sum across the row, cumulating the denominator for Equation (4.14). Then we pass across the row a second time, dividing by the sum to get this row's probabilities, which of course must sum to one.

If you are using the "correct" alpha and beta, rather than my normalized versions, the algorithm just shown will massively fail. When the log alphas and betas are exponentiated to get their actual values for summing the terms of Equation (4.13), it is virtually certain that for some or even all cases, all terms will be a machine zero, thus hitting any floor applied. In such cases, all normalized terms of Equation (4.13) will equal $1/N$, leading to nonsensical results.

In the file HMM.CPP, I show how to normalize to handle the "correct" case. I will not discuss that in this text, because I do not recommend it. But for educational purposes, I am going to now present this normalization in the context of my normalized alpha and beta. This is how I programmed it in HMM.CPP. Honestly, it's overkill, but it does illustrate proper normalization, and it perhaps slightly increases the accuracy of results. Here is the code, and I'll provide a brief explanation after.

```
for (istate=0 ; istate<nstates ; istate++) {
   for (jstate=0 ; jstate<nstates ; jstate++)
      trans_work2[istate*nstates+jstate] = 0.0 ;
   }

for (icase=1 ; icase<ncases ; icase++) {

   // Find max product for scaling; save each product to avoid recomputing
   for (istate=0 ; istate<nstates ; istate++) {
      for (jstate=0 ; jstate<nstates ; jstate++) {
         if (alpha[(icase-1)*nstates+istate] > exp(-300.0))
            temp = log ( alpha[(icase-1)*nstates+istate] ) ;
         else
            temp = -300.0 ;
         if (transition[istate*nstates+jstate] > exp(-300.0))
            temp += log ( transition[istate*nstates+jstate] ) ;
         else
            temp -= 300.0 ;
         if (densities[jstate*ncases+icase] > exp(-300.0))
            temp += log ( densities[jstate*ncases+icase] ) ;
         else
            temp -= 300.0 ;
         if (beta[icase*nstates+istate] > exp(-300.0))
            temp += log ( beta[icase*nstates+jstate] ) ;
         else
            temp -= 300.0 ;
         trans_work1[istate*nstates+jstate] = temp ;
         if (istate == 0 && jstate == 0)
            max_prod = temp ;
         else {
            if (temp > max_prod)
               max_prod = temp ;
            }
         } // For jstate
      } // For istate, finding max and saving log products
```

This is the first half of the code. As before, we begin by zeroing the matrix where we will cumulate the numerators of Equation (4.14). In the simple version of the algorithm, seen a few pages back, we summed the terms as the first step in the icase loop. But here we need a preliminary step. Instead of working with the actual terms, we work with their logs. To prevent accidentally taking the log of zero, we place an otherwise innocuous floor under each of the four terms individually and sum the logs (which is equivalent to multiplying the terms). We could alternatively multiply the terms and then take the log of the product, placing a floor under the product first. That would be almost equivalent to what is shown here, but I slightly prefer treating the terms individually. In practice it probably makes no difference which method is used. Then, this first step keeps track of the maximum sum-of-logs in max_prod.

```
// Sum the exp() of the shifted terms
sum = 0.0 ;
for (istate=0 ; istate<nstates ; istate++) {
  for (jstate=0 ; jstate<nstates ; jstate++) {
    temp = exp ( trans_work1[istate*nstates+jstate] - max_prod ) ;
    trans_work1[istate*nstates+jstate] = temp ;
    sum += temp ;
    } // For jstate
  } // For istate

// Normalize to probabilities
for (istate=0 ; istate<nstates ; istate++) {
  for (jstate=0 ; jstate<nstates ; jstate++) {
    temp = trans_work1[istate*nstates+jstate] ;
    trans_work2[istate*nstates+jstate] += temp / sum ;
    }
  }
 } // For icase

for (istate=0 ; istate<nstates ; istate++) {
  sum = 0.0 ;
  for (jstate=0 ; jstate<nstates ; jstate++) {
    temp = trans_work2[istate*nstates+jstate] ;
    if (temp < 1.e-200)
      temp = 1.e-200 ;
```

```
    sum += temp ;
    }
for (jstate=0 ; jstate<nstates ; jstate++) {
    temp = trans_work2[istate*nstates+jstate] ;
    if (temp < 1.e-200)
        temp = 1.e-200 ;
    transition[istate*nstates+jstate] = temp / sum ;
    }
}
```

The second of the three steps inside the icase loop is analogous to the first step in the simpler version. But now, instead of having just computed the actual product, we have saved in trans_work1 the log of the product. We subtract the maximum of all terms and exponentiate the difference. This way, the maximum term is one, and all other terms are scaled down accordingly. This gives us clean arithmetic. Each of these terms is saved right back onto where it came from, and the terms are summed.

The last of these three steps is analogous to the second step in the simpler version: we normalize to get an actual but rescaled value of an Equation (4.13) term and sum it into trans_work2, cumulating the numerator terms of Equation (4.14).

We complete the process in exactly the same way we complete the simpler version, first summing the numerators of Equation (4.14) across each row and then dividing to normalize each row to sum to 1. The floor used here is of little importance, as long as it is tiny. I chose to use a smaller floor than in the simple version just because we have more accuracy down there, but in practice it probably makes no difference whatsoever.

Assessing HMM Memory in a Time Series

On Page 142 we will see the ultimate goal of this chapter: linking measurable feature variables to an unmeasurable target variable by means of an underlying hidden Markov model. But it makes no sense doing that if our candidate features do not have memory that can be modeled by a hidden Markov model. Thus, our preliminary step should be to assess whether our feature variables, alone or in small groups, have memory that can be explained by a hidden Markov model.

Alternatively, and especially if we have an unwieldy quantity of candidate variables, we may wish to reverse this order: first, perform the linkage test, and then confirm that the selected features conform satisfactorily to a hidden Markov model explanation.

Of course, if they do not, the linkage test will often fail, and if they do, the linkage test will often succeed (if such linkage is actually present!). However, conflicts do arise and can be quite revealing. If the linkage test shows a strong relationship but the memory test described in this section shows a poor HMM explanation, we should be inclined to largely disregard the linkage results and focus on more traditional data mining techniques. Conversely, if the linkage test fails but the HMM test succeeds, we have pretty good evidence that the features have little predictive power for the target variable, stronger evidence than what could be obtained by most traditional tests alone. Thus, it behooves us to perform *both* tests, ideally but not necessarily doing the memory test first.

The code for performing the HMM memory test is heavily integrated with the *VarScreen* interface and database structure, so I won't present a complete subroutine. Rather, I'll provide an overview of the threaded version of the algorithm, along with occasional code snippets for illustration. This should enable most readers to program their own version of the algorithm. You can find all of these routines in the file HMM_MEM.CPP. That file also includes a wrapper module that will *not* compile correctly, because it is heavily integrated into the rest of the *VarScreen* program. However, that wrapper should serve as a solid template for using these routines in your own program. I also included in that module many debugging statements that you may find useful.

The user will supply the following:

- The number of predictor variables (npred) and cases (n_cases) and the ability to fetch them as well as the target variable

- The number of states to assume (n_states)

- The number of initialization trials (n_init)

- The maximum number of training iterations (max_iters)

- The number of Monte-Carlo replications (mcpt_reps)

A single data structure, a copy of which exists for each thread, takes care of parameter passing:

```
typedef struct {
   int irep ;        // Replication number (0 is unpermuted)
   HMM *hmm ;   // HMM object for this thread
   double *data ;  // Data to process, n_cases rows by npred columns
   double loglike ; // Computed log likelihood is returned here
} MEM_PARAMS ;
```

We have to allocate two work arrays as well as an HMM object for each thread. The first npred∗n_cases block in the pred array will be the "reference" dataset. It begins as the original data and is shuffled for Monte-Carlo replications. The remaining max_threads blocks will be private areas for the executing threads. As soon as pred is allocated, the original data should be copied to its first npred∗n_cases block.

```
pred = (double *) MALLOC ( (max_threads+1) * npred * n_cases * sizeof(double) ) ;
crits = (double *) MALLOC ( mcpt_reps * sizeof(double) ) ;

for (ithread=0 ; ithread<max_threads ; ithread++)
  mem_params[ithread].hmm = new HMM ( n_cases , npred , n_states ) ;
```

We initialize and start the main loop, which executes each Monte-Carlo permutation in a separate thread.

```
n_threads = 0 ;      // Counts threads that are active
irep = 0 ;           // MCPT replication number (a thread)
empty_slot = -1 ;    // After full, will identify the thread that just completed
for (;;) {           // Main thread loop processes all MCPT replications
```

If we are in a permutation pass (all but the first pass), shuffle the "reference" dataset at the beginning of pred.

```
if (irep) {                          // If doing permuted runs, shuffle
  i = n_cases ;                      // Number remaining to be shuffled
  while (i > 1) {                    // While at least 2 left to shuffle
    j = (int) (unifrand () * i) ;    // MUST be thread-safe; threads are running
    if (j >= i)                      // Should never happen, but be safe
      j = i - 1 ;
    --i ;
    for (k=0 ; k<npred ; k++) {
      dtemp = pred[i*npred+k] ;
      pred[i*npred+k] = pred[j*npred+k] ;
      pred[j*npred+k] = dtemp ;
      }
    }
  } // If in permutation run (irep > 0)
```

Insert a check here for the user pressing the ESCape key. That code is omitted for clarity. Then start a new thread if there is still work to do.

```
if (irep < mcpt_reps) { // If there are still some to do
  if (empty_slot < 0)   // Negative while we are initially filling the thread queue
    k = n_threads ;
  else
    k = empty_slot ;
  dptr = pred + (k+1) * npred * n_cases ; // This thread's data goes here
  memcpy ( dptr , pred , npred * n_cases * sizeof(double) ) ;
  mem_params[k].data = dptr ;
  mem_params[k].irep = irep ;
  threads[k] = (HANDLE) _beginthreadex ( ... ) ;
  ++n_threads ;
  ++irep ;
  } // if (irep < mcpt_reps)
```

Check to see if we are done, and break if so. Otherwise, handle the situation of the full suite of threads running and there are more threads to add as soon as some are done. We sit here and wait for just one thread to finish.

```
if (n_threads == 0) // Are we done?
  break ;

if (n_threads == max_threads && irep < mcpt_reps) { // Full suite but more to do?
  ret_val = WaitForMultipleObjects ( n_threads , threads , FALSE , INFINITE ) ;
  // Check here for error return...
  crits[mem_params[ret_val].irep] = mem_params[ret_val].loglike ;
  empty_slot = ret_val ;     // This slot can now accept a new thread
  CloseHandle ( threads[empty_slot] ) ;
  threads[empty_slot] = NULL ;
  --n_threads ;
  }
```

Handle the situation that all work has been started and now we are just waiting for threads to finish.

```
else if (irep == mcpt_reps) {
  ret_val = WaitForMultipleObjects ( n_threads , threads , TRUE , INFINITE ) ;
  // Check here for error return...
  for (i=0 ; i<n_threads ; i++) {
    crits[mem_params[i].irep] = mem_params[i].loglike ;
    CloseHandle ( threads[i] ) ;
    }
  break ;
  }

} // For all MCPT reps
```

Most work is done. Now we count the number of permuted replications whose log likelihood equaled or exceeded that of the original data.

```
original = crits[0] ;
count = 1 ;
for (irep=1 ; irep<mcpt_reps ; irep++) {
  if (crits[irep] >= original)
    ++count ;
}
```

This count gives us a simple Monte-Carlo permutation test for the null hypothesis that the data cannot be explained by a hidden Markov model. If this null hypothesis is true (the data has no HMM memory), we would expect that the log likelihood of the original data would be about the same as those of the permuted datasets, which by definition have no HMM memory. But if the data is well fitted by a hidden Markov model, we would expect its log likelihood to be greater than that of most or all of the permuted datasets, leading to a very small count. In fact, count/mcpt_reps is the probability that, if the null hypothesis is true (the data has no HMM memory), we could have gotten a log likelihood as great as we observed by pure luck. When we perform this test, we really want a probability no greater than 0.05, and a cutoff of 0.01 is nicely conservative.

Linking Features to a Target

At last we come to the ultimate goal of this chapter. Rather than taking the traditional approach of assuming that predictor variables (features) have a direct, perhaps even causal, relationship with a target, we assume the existence of a hidden underlying process that is always in one of several possible states, and whose transition probability from one state to another depends on what state it is transitioning from. Each state implies a different probability distribution of associated variables, some of which are measurable (our predictors) and at least one of which is not (our target). We use the measured variables to make educated guesses for the state of the process as of each observation, and then knowledge of the state probabilities lets us make an educated guess for the target.

The algorithm takes place in three distinct steps. *In the first step we ignore the target.* This concept may be foreign to people new to this approach. The first step analyzes the time series of one or more features and finds a hidden Markov model that best fits the data. At this point, to save computation time, *VarScreen* does not try to determine whether the model is a good fit. That's why the memory test described in the prior section is important. All that the first step gives us is the best model, good or not.

The second step relates the state to the target. This is done by using ordinary linear regression, letting the vector of state probabilities for each case predict the target for that case. This is more powerful than just identifying the most likely state and relating that nominal variable to the target, because it also considers the confidence in the most likely state, as well as the confidences in the alternative, less likely states. The multiple-R of this regression serves as the relationship criterion.

The third step applies a Monte-Carlo permutation test to the relationship just found, letting us compute a p-value for the null hypothesis that there is no relationship between the process state and the target. Unfortunately, there is no way to perform a perfect MCPT, due to the fact that both the predictors and (almost certainly) the target have serial correlation. So instead of the ideal complete permutation, we use cyclic permutation, which largely (though not completely) preserves serial correlation. This results in p-values with larger error variance, an unavoidable annoyance.

VarScreen adds a nice extension to the steps just stated. Instead of fitting a single small set of predictors to an HMM, the user can specify a larger collection of predictor candidates as well as the number of predictors in the model (the *dimension*, 1-3). The program then computes an optimal HMM for every possible set of predictors taken *dimension* at a time and prints for the user the parameters of the set whose HMM has greatest multiple-R with the target. It also computes MCPT p-values that are compensated for the multiple comparisons inherent in this action. Here is a rough outline of the procedure. More details and code snippets will follow.

1) For all combinations of candidates, taken *dimension* at a time...

   2) If there are more combinations to test, assign a thread to fit an HMM to this combination.

   3) If all threads are running and there are more combinations to test, wait for a thread to finish, save the HMM, and go to step 2.

   4) If all combinations have been assigned to a thread, wait for all threads to finish. Save the HMMs and exit this loop.

5) For all MCPT replications...

   6) If we are past the first replication, shuffle the target.

   7) For all candidate combination HMMs...

     8) Compute the criterion relating the HMM to the (possibly shuffled) target. Keep track of the *best criterion* of all HMMs.

     9) If this is the first (unshuffled) MCPT replication, save the *original criterion* for this HMM and initialize the solo and unbiased MCPT counts to 1 for this HMM.

    10) If this is not the first MCPT replication, and if this (shuffled) criterion equals or exceeds the unshuffled criterion for this HMM, increment the solo MCPT count for this HMM.

  11) If we are in the first (unshuffled) MCPT replication, sort the HMM criteria saved in step 9 for printing later.

  12) If we are not in the first MCPT replication, then for all HMMs...

    13) If the *best criterion* saved in step 8 equals or exceeds the *original criterion* saved in step 9, increment the unbiased MCPT counter for this HMM.

Before proceeding to details, we take an intuitive view of the algorithm just shown. The loop in step 1, which comprises steps 2–4, fits and saves an HMM for every possible combination of *dimension* predictors in the user-supplied candidate list. Each combination is handled by a single thread. This is by far the most time-consuming part of the complete algorithm.

The loop of step 5 handles all MCPT replications, the first of which is the original, unshuffled target set. In step 7 we pass through all HMMs computed and saved in steps 1–4. For each, compute the linear regression criterion relating the HMM to the target (which will be shuffled after the first MCPT replication). Keep track of the best (largest criterion) HMM, understanding that this *best criterion* is for the current MCPT replication, so after the first replication it will be referring to shuffled targets.

If this is the first (unshuffled) MCPT replication, we are dealing with actual target data, so we need to preserve the original criterion for each HMM. We'll want to print and otherwise refer to this information later. But if we are into a shuffled replication, then we compare this criterion (for each HMM) to the saved original (unshuffled) criterion. If, to our dismay, this criterion that was computed from shuffled targets is at least as good as the criterion obtained from the unshuffled targets, we increment the solo MCPT counter. Later, when all replications are finished, we can divide this count by the total number of MCPT replications to compute the p-value for the null hypothesis that this HMM, when considered in isolation (solo), has no relationship with the target.

After processing all HMMs for this MCPT replication, we reach step 11. If this is the first (unshuffled) replication, we sort the HMMs' criteria so that later we can print the HMMs in order of relationship to the target. No matter which replication we are in, we saved (in step 8) the best criterion across all HMMs. So if we are in a shuffled replication, we compare this replication's best criterion to all original criteria, incrementing the unbiased MCPT counter whenever this shuffled best equals or exceeds the original value. Note that for whichever of these original criteria was the best, we will be comparing apples to apples, or in this case, best to best. Thus, for the best original HMM, we can divide its count by the number of replications to get a p-value for the null hypothesis that none of the HMMs has any relationship with the target.

Now for some details and code snippets. The user specifies some items:

> max_threads – Maximum number of threads to use
>
> ncases – Number of cases
>
> nDim – Dimension of HMM, 1-3
>
> nstates – Number of states

nX – Number of predictor candidates, at least nDim

Xindices – Their database indices are here

We also have a data structure for passing parameters for each thread that trains an HMM:

```
typedef struct {
  HMM *hmm ;       // The HMM for this thread
  int ncases ;     // Number of cases in data
  int ncols ;      // Number of columns in data
  double *data ;   // All data, used and unused, is here; points to global 'database'
  int nDim ;       // Dimension of HMM, 1-3
  int nstates ;    // Number of states
  int icombo ;     // Index of predictor combination needed for placing result
  int pred1 ;      // Index in database of first predictor; always used
  int pred2 ;      // And second predictor; used if nDim >= 2
  int pred3 ;      // And third predictor; used if nDim == 3
  double *datawork ; // Work area ncases * nDim long
  double crit ;    // Returns log likelihood
} HMM_PARAMS ;
```

In this structure, icombo is the set of predictor indices pred1-pred3 encoded as a single integer. Thus, the former integer and the latter set are redundant. But it will often be handy to have both available for different purposes, thus avoiding the need to encode or decode.

We also have a class dedicated to passing learned HMM parameters:

```
class HMMresult {

public:
  HMMresult ( int ncases , int nDim , int nstates ) ;
  ~HMMresult () ;

  int ok ;         // Did memory allocation go well?
  int pred1 ;      // Index in database (not preds) of first predictor
  int pred2 ;      // And second
  int pred3 ;      // And third
```

```
  double *means ;         // Nstates * nDim means; state changes slowest
  double *covars ;        // Nstates * nDim * nDim covariances; state changes slowest
  double *init_probs ;    // Nstates vector of probability that first case is in each state
  double *transition ;    // Nstates * nstates transition probability matrix; Aij=prob(i-->j)
  double *state_probs ;   // Ncases * nstates probabilities of state
} ;
```

We need a set of thread parameters and thread handles for each thread. First, we initialize the thread parameters that never change. These could be global, but it's cleaner programming doing it this way.

```
HMM_PARAMS hmm_params[MAX_THREADS] ;
HANDLE threads[MAX_THREADS] ;

for (ithread=0 ; ithread<max_threads ; ithread++) {
  hmm_params[ithread].hmm = new HMM ( n_cases , nDim , nstates ) ;
  hmm_params[ithread].ncases = n_cases ;
  hmm_params[ithread].ncols = n_vars ;
  hmm_params[ithread].data = database ;
  hmm_params[ithread].nDim = nDim ;
  hmm_params[ithread].nstates = nstates ;
  hmm_params[ithread].datawork = datawork + ithread * ncases * nDim ;
  } // For all threads, initializing constant stuff
```

We compute the total number of combinations of nX things taken nDim at a time that we will need to process. Also initialize the first combination.

```
ipred1 = icombo = 0 ;
n_combo = nX ;

if (nDim > 1) {
  ipred2 = 1 ;
  n_combo = n_combo * (nX-1) / 2 ;

  if (nDim > 2) {
    ipred3 = 2 ;
    n_combo = n_combo * (nX-2) / 3 ;
    }
  }
```

We have to allocate the hmm_results array:

```
hmm_results = (HMMresult **) MALLOC ( n_combo * sizeof(HMMresult *) ) ;
for (i=0 ; i<n_combo ; i++) {
  hmm_results[i] = new HMMresult ( n_cases , nDim , nstates ) ;
```

Get ready to start the main loop that sets threads to work processing each combination of predictor candidates.

```
n_threads = 0 ;              // Counts threads that are active
for (i=0 ; i<max_threads ; i++)
  threads[i] = NULL ;

title_progress_message ( "Starting initial suite of threads..." ) ;
setpos_progress_message ( 0.0 ) ;

empty_slot = -1 ; // Flags still filling; After full, identifies the thread that just completed
for (;;) {       // Main thread loop processes all combinations
```

This loop will continue assigning combinations of candidates to threads until all are complete. It's polite to see if the user has pressed the ESCape key. Because we do this check before refilling an empty, completed thread, we need to compress all running threads to the beginning of the thread handle array before waiting for them to finish.

```
if (escape_key_pressed || user_pressed_escape ()) {
  for (i=0, k=0 ; i<max_threads ; i++) { // Compress to start of array
    if (threads[i] != NULL)
      threads[k++] = threads[i] ;
    }
  ret_val = WaitForMultipleObjects ( k , threads , TRUE , INFINITE ) ;
  // Would be nice to check for unexpected error return here
  for (i=0 ; i<k ; i++) {
    CloseHandle ( threads[i] ) ;
    threads[i] = NULL ;        // Not really needed
    }
  return ERROR_ESCAPE ;
  }
```

This next block of code is a little tricky. We see if there are still more combinations to try. If so, we put the essential parameters for this combination in the parameter-passing structure and start a new thread. Then we advance to the next combination of predictor candidates. If we are still filling the initial queue of threads, empty_slot will still be negative. But after it's full, whenever a thread finishes its task, its index will be assigned to empty_slot, which will be immediately filled here.

```
if (icombo < n_combo) {       // If there are still some combinations to do
  if (empty_slot < 0)          // Negative while we are initially filling the queue
    k = n_threads ;
  else
    k = empty_slot ;
  hmm_params[k].icombo = icombo ;   // Needed for placing final result
  hmm_params[k].pred1 = Xindices[ipred1] ;
  if (nDim > 1)
    hmm_params[k].pred2 = Xindices[ipred2] ;
  if (nDim > 2)
    hmm_params[k].pred3 = Xindices[ipred3] ;
  threads[k] = (HANDLE) _beginthreadex ( ... ) ;
  ++n_threads ;

  // Advance to the next combination
  ++icombo ;
  ++ipred1 ;
  if (nDim > 1) {
    if (ipred1 == ipred2) {
      ++ipred2 ;
      ipred1 = 0 ;
      }
    if (nDim > 2) {
      if (ipred2 == ipred3) {
        ++ipred3 ;
        ipred1 = 0 ;
        ipred2 = 1 ;
        }
      }
    }
```

```
} // if (icombo < n_combo) (Another combination left to do)

if (n_threads == 0) // Are we done?
  break ;
```

The next block of code handles the situation of the full suite of threads running and there are more threads (combinations) to add as soon as some are done. We sit here and wait for just one thread to finish. Then we copy the learned parameters for this HMM to the hmm_results object for this combination. This thread slot is now available for reuse, so we copy it to empty_slot so that the next thread goes there. Of course we need to close the thread to release its resources to the system, and we decrement the counter of the number of threads running.

```
if (n_threads == max_threads && icombo < n_combo) {
  ret_val = WaitForMultipleObjects ( n_threads , threads , FALSE , INFINITE ) ;
  // You really should check for an unexpected error return here
  // Copy the results of this run to the output vector hmm_results
  k = hmm_params[ret_val].icombo ; // Result goes here in output vector
  memcpy ( hmm_results[k]->means ,
          hmm_params[ret_val].hmm->means ,
          nDim*nstates*sizeof(double) ) ;
  memcpy ( hmm_results[k]->covars ,
          hmm_params[ret_val].hmm->covars ,
          nDim*nDim*nstates*sizeof(double) ) ;
  memcpy ( hmm_results[k]->init_probs ,
          hmm_params[ret_val].hmm->init_probs ,
          nstates*sizeof(double) ) ;
  memcpy ( hmm_results[k]->transition ,
          hmm_params[ret_val].hmm->transition ,
          nstates*nstates*sizeof(double) ) ;
  memcpy ( hmm_results[k]->state_probs ,
          hmm_params[ret_val].hmm->state_probs ,
          nstates*ncases*sizeof(double) ) ;
  hmm_results[k]->pred1 = hmm_params[ret_val].pred1 ;
  hmm_results[k]->pred2 = hmm_params[ret_val].pred2 ;
  hmm_results[k]->pred3 = hmm_params[ret_val].pred3 ;
```

```
      empty_slot = ret_val ;
      CloseHandle ( threads[empty_slot] ) ;
      threads[empty_slot] = NULL ;
      --n_threads ;
      }
```

The last major block of code in the thread loop handles the situation that all combinations have been submitted to threads, and now we are just waiting for them to finish. As soon as they all finish, we pass through all of them and collect their results.

```
   else if (icombo == n_combo) {
     ret_val = WaitForMultipleObjects ( n_threads , threads , TRUE , INFINITE ) ;
     // Should check for an error return here and act accordingly

     for (i=0 ; i<n_threads ; i++) {
       k = hmm_params[i].icombo ; // Result goes here in output vector
       memcpy ( hmm_results[k]->means ,
                hmm_params[i].hmm_core->means ,
                nDim*nstates*sizeof(double) ) ;
       memcpy ( hmm_results[k]->covars ,
                hmm_params[i].hmm_core->covars ,
                nDim*nDim*nstates*sizeof(double) ) ;
       memcpy ( hmm_results[k]->init_probs ,
                hmm_params[i].hmm_core->init_probs ,
                nstates*sizeof(double) ) ;
       memcpy ( hmm_results[k]->transition ,
                hmm_params[i].hmm_core->transition ,
                nstates*nstates*sizeof(double) ) ;
       memcpy ( hmm_results[k]->state_probs ,
                hmm_params[i].hmm_core->state_probs ,
                nstates*ncases*sizeof(double) ) ;
       hmm_results[k]->pred1 = hmm_params[i].pred1 ;
       hmm_results[k]->pred2 = hmm_params[i].pred2 ;
       hmm_results[k]->pred3 = hmm_params[i].pred3 ;
       CloseHandle ( threads[i] ) ;
       }
```

```
      break ;
    }
} // Endless loop which threads computation of criterion for all predictors
```

That completes the first stage of the process. We have computed and saved the parameters of an HMM for every possible combination of predictor candidates. We have not saved the associated likelihoods because these are irrelevant to the task. For economy, we also have avoided showing the code for fetching the predictors and calling hmm::estimate(), as that was covered in the *HMM Memory* section.

Linking HMM States to the Target

Now that we have an HMM for every possible combination of predictor candidates, we can proceed with the second phase of the operation, evaluating for each HMM the association between its state and the target variable. At this point we have available, for each case, the vector of computed probabilities that the case is in each of the possible states. We also have the target value for that case.

The naive way to evaluate the association between state and target would be to classify each case as being in whichever class has the largest state probability and then finding the mean target value for each class. There are two problems with this approach. First, if we plan to use this association method to make future predictions (which we do not do here, but which is easily done because we know the learned HMM parameters), our predictions would have only a few discrete values, one for each state. For a continuous variable, this is clearly unacceptable.

More significantly, this naive approach ignores the probability of being in states other than that having the largest probability. This is important information that we should not ignore. My algorithm takes this into account by using ordinary linear regression to build a linear model that uses the complete set of state probabilities to predict the target. Not only does this take advantage of all available state information, but it has the bonus of giving us a numerical indication of the way state membership relates to the target, if we are interested. In the *VarScreen* program, I do not print the regression coefficients, choosing instead to print the correlations, because these tend to be more stable in the presence of significant correlation among predictors. However, the coefficients are readily available if you wish to examine them. We do the linear regression with singular value decomposition, so we need to allocate that object:

```
SingularValueDecomp *sptr ;
sptr = new SingularValueDecomp ( n_cases , nstates+1 , 0 ) ;
```

It's easiest to include the Monte-Carlo permutation test into the computation of associations. As is my habit, I let the first pass through the loop handle the unpermuted data and permute thereafter. Here is the first section of this code, that which handles shuffling the target.

```
for (irep=0 ; irep<mcpt_reps ; irep++) {
  if (irep) {                  // If doing permuted runs, shuffle

    if (mcpt_type == 1) { // Complete version of shuffling
      i = n_cases ;          // Number remaining to be shuffled
      while (i > 1) {        // While at least 2 left to shuffle
        j = (int) (unifrand_fast () * i) ;   // Not thread-safe routine, which is okay here
        if (j >= i)
          j = i - 1 ;
        --i ;
        dtemp = datawork[i] ; // The target is here
        datawork[i] = datawork[j] ;
        datawork[j] = dtemp ;
        }
      } // Type 1, complete

    else if (mcpt_type == 2) { // Cyclic
      j = (int) (unifrand_fast () * n_cases) ;
      if (j >= n_cases)
        j = n_cases - 1 ;
      for (i=0 ; i<n_cases ; i++)
        datawork[n_cases+i] = datawork[(i+j)%n_cases] ;
      for (i=0 ; i<n_cases ; i++)
        datawork[i] = datawork[n_cases+i] ;
      } // Type 2, cyclic

    } // If in permutation run (irep > 0)
```

The preceding code handles the shuffling, which may be of the complete type or the cyclic type, which must be used if the target has significant serial correlation. (The predictors are almost certainly serially correlated.) Prior to executing this code, datawork was allocated to be at least 2*n_cases long, so we can use the later set of n_cases slots as temporary storage. The cyclic shuffling algorithm shifts (with endpoint wraparound) the target array, copying the shifted elements to this storage area, and then it just copies it back. Cyclic shuffling in place is more complex and not significantly faster.

The SingularValueDecomp object does not destroy its right-hand-side inputs, so we can place the target vector in its b member once now and leave it there as we test the various models. (The file SVDCMP.CPP contains detailed instructions for using this object to perform linear regression. Please refer to it if needed.) Also, compute the variance of the target to enable computation of multiple-R criteria later.

```
memcpy ( sptr->b , datawork , n_cases * sizeof(double) ) ;   // Get the target

mean = var = 0.0 ;
for (i=0 ; i<n_cases ; i++)
  mean += datawork[i] ;
mean /= n_cases ;

for (i=0 ; i<n_cases ; i++) {
  diff = datawork[i] - mean ;
  var += diff * diff ;
  }
var /= n_cases ;
```

We now pass through all of the previously computed HMMs. For each we compute its multiple-R criterion by fitting the linear model. The state_probs array in the hmm_results object for this predictor combination is already arranged almost exactly as needed for sptr->a, the matrix of independent variables. All we need to do is include the constant 1.0 as the constant term in the linear equation. Then we solve the singular value decomposition and back-substitute to get the regression coefficients.

```
for (icombo=0 ; icombo<n_combo ; icombo++) {
  // You should check for the user pressing ESCape here

  aptr = sptr->a ;
  dptr = hmm_results[icombo]->state_probs ; // Independent vars in linear equation
```

153

```
    for (i=0 ; i<n_cases ; i++) {
      for (j=0 ; j<nstates ; j++)
        *aptr++ = dptr[i*nstates+j] ;
      *aptr++ = 1.0 ; // Constant term
      }
    sptr->svdcmp () ;
    sptr->backsub ( 1.e-7 , coefs ) ; // Computes optimal weights
```

We now have the linear regression coefficients. Pass through all cases and cumulate the squared error. From this we can easily compute the R-square, which is the square of the multiple-R. Either can be used as our relationship criterion, as they are monotonically related. In unusual pathological situations, it may be that the prediction is worse than worthless, in which case R-square will be negative. Since we will later be printing its square root, the multiple-R, we bound it at zero, which also improves the MCPT in such bizarre (and very rare!) situations. Keep track of the best HMM criterion (best_crit) for this MCPT replication.

If this is the first (unpermuted) replication, save the criterion for each combination and initialize the solo and unbiased MCPT counters. But if this is a permuted replication, update the solo counter as appropriate.

```
  error = 0.0 ;
  for (i=0 ; i<n_cases ; i++) {
    sum = coefs[nstates] ;        // Constant term
    for (j=0 ; j<nstates ; j++)    // Compute prediction of target from state probs
      sum += dptr[i*nstates+j] * coefs[j] ;
    diff = sum - datawork[i] ;    // Predicted minus actual
    error += diff * diff ;
    }
  error /= n_cases ;             // MSE

  crit = 1.0 - error / var ;      // R-squared (square of multiple-R)
  if (crit < 0.0)
    crit = 0.0 ;                  // We'll be printing square-root, so must not be negative

  if (icombo == 0 || crit > best_crit)
    best_crit = crit ;
```

```
if (irep == 0) {                        // Original, unpermuted data
   sorted_crits[icombo] = original_crits[icombo] = crit ;
   index[icombo] = icombo ;
   mcpt_bestof[icombo] = mcpt_solo[icombo] = 1 ;
   }

else if (crit >= original_crits[icombo])
   ++mcpt_solo[icombo] ;

} // For all HMM models (icombo)
```

We're almost finished with all computation. If this is the first (unpermuted) replication, sort the true criteria ascending, simultaneously moving their indices so we can later identify which is which. But if this is a permuted replication, update the unbiased MCPT counter. For each combination, increment the counter if the best criterion in this replication equaled or exceeded the combination's criterion. For whichever original criterion was the greatest among all of the original criteria, this will be an apples-to-apples comparison (best-to-best here), so the counter divided by mcpt_reps will be a true p-value for the null hypothesis that, if all combinations were worthless, the best could have obtained its exalted criterion by pure luck. But for combinations below the best, this ratio will be an upper bound on the true but unknown p-value.

```
if (irep == 0) { // Get the indices that sort the HMM models' criterion with target
   qsortdsi ( 0 , n_combo-1 , sorted_crits , index ) ;
   ibest = index[n_combo-1] ;    // We'll print specs of this model
   }

else {
   for (icombo=0 ; icombo<n_combo ; icombo++) {
      if (best_crit >= original_crits[icombo]) // Valid for only the largest
         ++mcpt_bestof[icombo] ;
      }
   }

} // For all MCPT replications
```

The code snippets just presented enable the reader to compute everything needed to perform an HMM–target relationship analysis. Things like memory allocation/deallocation and error handling have been omitted for clarity but should be easily handled in a manner consistent with the programmer's style. The same could be said for printing results, but I will show how I chose to present results to the user. You may wish to use my method as your template or modify it to your own liking.

We print the means and standard deviations of the variables that define the best (highest multiple-R with the target) HMM. Separate blocks of code are needed for the cases of having one, two, or three variables in the model. I'll show just the code for two and three variables; that for one variable should be obvious.

```
best_result = hmm_results[ibest] ;      // This is the highest-correlating model
sprintf ( msg, "Specifications of the best HMM model correlating with %s...",
          var_names[target] ) ;
audit ( msg ) ;
audit ( "Means (top number) and standard deviations (bottom number)" ) ;

if (nDim == 1) {
  ...
  }
else if (nDim == 2) {
  sprintf ( msg, "State %15s %15s",
            var_names[best_result->pred1], var_names[best_result->pred2] ) ;
  audit ( msg ) ;
  for (i=0 ; i<nstates ; i++) {
    sprintf ( msg, "%4d %12.5lf %16.5lf", i+1,
              best_result->means[2*i], best_result->means[2*i+1] ) ;
    audit ( msg ) ;
    sprintf ( msg, " %12.5lf %16.5lf",
              sqrt ( best_result->covars[4*i] ), sqrt ( best_result->covars[4*i+3] )) ;
    audit ( msg ) ;
    }
  }
else if (nDim == 3) {
  sprintf ( msg, "State %15s %15s %15s", var_names[best_result->pred1],
            var_names[best_result->pred2], var_names[best_result->pred3] ) ;
  audit ( msg ) ;
  for (i=0 ; i<nstates ; i++) {
    sprintf ( msg, "%4d %12.5lf %16.5lf %16.5lf", i+1, best_result->means[3*i],
              best_result->means[3*i+1], best_result->means[3*i+2] ) ;
```

```
    audit ( msg ) ;
    sprintf ( msg, " %12.5lf %16.5lf %16.5lf", sqrt ( best_result->covars[9*i] ),
            sqrt ( best_result->covars[9*i+4] ), sqrt ( best_result->covars[9*i+8] )) ;
    audit ( msg ) ;
    }
}
```

We also print the transition probability matrix of this best HMM.

```
audit ( "Transition probabilities..." ) ;
audit ( "" ) ;

sprintf ( msg, "    %3d", 1 ) ;
for (i=2 ; i<=nstates ; i++) {
  sprintf ( msg2, "    %3d", i ) ;
  strcat ( msg , msg2 ) ;
  }
audit ( msg ) ;

for (i=0 ; i<nstates ; i++) {
  sprintf ( msg, "%3d", i+1 ) ;
  for (j=0 ; j<nstates ; j++) {
    sprintf ( msg2, "%9.4lf", best_result->transition[i*nstates+j] ) ;
    strcat ( msg , msg2 ) ;
    }
  audit ( msg ) ;
  }
```

It is nice to print a table of other interesting information. I devote a row to each state and print the following information across columns. I'll say a little more about each of these items later when examples are presented.

- Percent of cases in which this state has the highest probability

- Correlation of this state's probability with the target

- Mean of the target when the process is in this state

- Standard deviation of the target when the process is in this state

I print a table header as follows. The mean of the target is also computed before continuing, because we will need it for computing correlations between state probabilities and the target. The code that loops over all states appears on the next page, and a discussion follows.

```
audit ( "State   Percent   Correlation   Target mean   Target StdDev" ) ;

ymean = 0.0 ;
for (i=0 ; i<n_cases ; i++)
   ymean += datawork[i] ;
ymean /= n_cases ;

for (i=0 ; i<nstates ; i++) {
  xmean = sum_xx = sum_yy = sum_xy = target_mean = target_ss = 0.0 ;
  for (j=0 ; j<n_cases ; j++)
    xmean += best_result->state_probs[j*nstates+i] ;
  xmean /= n_cases ;    // Mean probability for this state
  win_count = 0 ;        // Counts times this state has max probability
  for (j=0 ; j<n_cases ; j++) {
    x = best_result->state_probs[j*nstates+i] ;   // This case's probability for this state
    y = datawork[j] ;                              // Target for this case
    xdiff = x - xmean ;
    ydiff = y - ymean ;
    sum_xx += xdiff * xdiff ;
    sum_yy += ydiff * ydiff ;
    sum_xy += xdiff * ydiff ;
    for (k=0 ; k<nstates ; k++) { // See if any other state probability >= this one
      if (k != i &&
        best_result->state_probs[j*nstates+i] <= best_result->state_probs[j*nstates+k])
        break ;
      }
    if (k == nstates) { // If this state beat (not just tied) all others
      ++win_count ;
      target_mean += y ;
      target_ss += y * y ;
      }
    }
```

```
sum_xx /= n_cases ;    // Not needed here because n_cases cancels below
sum_yy /= n_cases ;    // But I do it in case these are otherwise needed
sum_xy /= n_cases ;
if (win_count > 0) {
    target_mean /= win_count ;
    target_ss /= win_count ;
    }
else {
    target_mean = 0.0 ;
    target_ss = 0.0 ;
    }

sprintf ( msg, "%3d %9.2lf %9.5lf %12.5lf %13.5lf",
    i+1, 100.0 * win_count / n_cases, sum_xy / (sqrt ( sum_xx * sum_yy ) + 1.e-20),
    target_mean, sqrt(target_ss - target_mean * target_mean) ) ;
audit ( msg ) ;
}
```

Each state is processed individually. We immediately zero terms that will be cumulated across cases. Compute the mean probability for this state and zero the counter of how many times this state is the winner.

The next eight lines cumulate the sums of squares and the cross product that we need for computing the correlation between this state probability and the target.

The loop over k sees if any other state has a probability that equals or exceeds the probability of this state for this case. If this never happens, this state is the winner for this case, in which case the if statement evaluates true, we tally the win, and we cumulate the mean and sum-squares for this special situation of this state being the winner.

After the cases are all processed, we divide by n_cases to get the mean squares and cross product. This division is not necessary for how these quantities are used here, because n_cases will cancel when the correlation is computed. However, I explicitly show it here in case the sum-squares are ever used for some other purpose.

Also, as long as we had at least one case for which this state was a winner, we divide by the count to get the mean and variance. When we compute and print the standard deviation, we use the "easy" direct formula. This formula is not recommended for general use because, if the mean is large relative to the standard deviation, the subtraction of the squared mean from the sum-squares can result in significant loss of precision. However, just printing this quantity is not a critical use, and it is unlikely that the mean will swamp out the standard deviation in any application for which this technique is applied. If this is a bad assumption, the more complex "individual difference" formula should be used, as was done when computing the correlations. That approach here would entail saving the winning cases for a second pass, a bit of a nuisance for little return.

The final step is to print the most highly correlated models' relationships with the target, sorted with the best first. We print the predictor variables that define the HMM, the state probabilities' multiple-R with the target, as well as the solo and unbiased p-values.

```
sprintf ( msg, "-------------> Hidden Markov Models correlating with %s <-------------",
      var_names[target] ) ;
   audit ( msg ) ;

 if (nDim == 1)
   audit ( " Predictor Multiple-R Solo pval Unbiased pval" ) ;
 else if (nDim == 2)
   audit ( " Predictor 1 Predictor 2 Multiple-R Solo pval Unbiased pval" ) ;
 else if (nDim == 3)
   audit ( " Predictor 1 Predictor 2 Predictor 3 Multiple-R Solo pval Unbiased" ) ;

 for (i=n_combo-1 ; i>=0 ; i--) { // They were sorted in ascending order, so print reverse

   if (max_printed-- <= 0)      // This user parameter limits the number printed
     break ;

   k = index[i] ;                // Get the pre-sorting index of this HMM

   if (nDim == 1)
     sprintf ( msg, "%15s %12.4lf",
           var_names[hmm_results[k]->pred1], sqrt(original_crits[k]) ) ;

   else if (nDim == 2)
     sprintf ( msg, "%15s %15s %12.4lf",
           var_names[hmm_results[k]->pred1],
           var_names[hmm_results[k]->pred2], sqrt(original_crits[k]) ) ;
```

```
else if (nDim == 3)
  sprintf ( msg, "%15s %15s %15s %12.4lf",
          var_names[hmm_results[k]->pred1],
          var_names[hmm_results[k]->pred2],
          var_names[hmm_results[k]->pred3], sqrt(original_crits[k]) ) ;

sprintf ( msg2, " %12.4lf %12.4lf",
          (double) mcpt_solo[k] / (double) mcpt_reps,
          (double) mcpt_bestof[k] / (double) mcpt_reps ) ;
strcat ( msg , msg2 ) ;
audit ( msg ) ;
} // For all of the best HMMs
```

The preceding code snippets are in the file HMM_LINK.CPP.

A Contrived and Inappropriate Example

This section demonstrates a revealing example of the algorithm using synthetic data to clarify the presentation. We will also explore what happens when we attempt to use this technique on data that is not well fit by a hidden Markov model. The variables in the dataset are as follows:

> *RAND0–RAND9* are independent (within themselves and with each other) random time series. These are the predictor candidates.
>
> *SUM12* = *RAND1* + *RAND2*. This is the target variable.

I chose to use two predictors and allow four states in the models. The program fits a hidden Markov model to each of the $(10*9)/2=45$ pairs of predictor candidates. Not surprisingly, the model based on RAND1 and RAND2 has the highest correlation with SUM12. Its means and standard deviations for each state are printed first:

Means (top number) and standard deviations (bottom number)

State	RAND1	RAND2
1	0.06834	-0.66014
	0.48729	0.21358
2	-0.73466	0.07687
	0.17187	0.54038
3	-0.02272	0.35902
	0.39033	0.39555
4	0.73542	0.08884
	0.17546	0.52133

RAND1 and RAND2 are totally random (they exist in only one state), so attempting to fit a hidden Markov model to them should be extremely unstable. Indeed, in ten runs of this test, twice the program found solutions in which the means of the states were all nearly zero, indicating no differentiation between states. But most of the time, it came up with a pattern essentially identical to the preceding one. This solution is remarkably similar to a sort of principal components decomposition: RAND1 distinguishes between State 2 and State 4, while RAND2 distinguishes between State 1 and State 3. Thus, knowledge of which of the four states the process is in provides great information about SUM12.

Next we see the transition probabilities. The figure in Row i and Column j is the probability that the process will transition from State i to State j. Not surprisingly, they are almost all identical. The relatively small discrepancies are just due to random variation in the data.

Transition probabilities...

	1	2	3	4
1	0.2638	0.2037	0.3494	0.1830
2	0.2438	0.1945	0.3638	0.1979
3	0.2130	0.1682	0.4174	0.2014
4	0.2404	0.2148	0.3272	0.2176

Further properties of each state are then printed:

Percent is the percentage of cases in which this state has the highest probability. The sum of these quantities across all states may not reach 100 percent, because cases in which there is a tie for the highest probability are not counted. If the data is continuous, this should almost never happen.

Correlation is the ordinary correlation coefficient between the target and the membership probability for this state. On first consideration, it might be thought that the beta weight in the linear equation predicting the target from the state probabilities would be the better quantity to print. But the beta weight is not printed at all due to the fact that such weights are notoriously unstable and hence uninformative. Suppose there is very high correlation between the membership probabilities of two states, a situation which is especially likely to happen if the user specifies more states than actually exist in the process. Then both of these probabilities could be highly correlated with the target, while they might actually have opposite signs for their beta weights!

Target mean is the mean of the target when this state has the highest membership probability. Cases in which there is a tie for maximum (almost impossible for continuous data) do not enter into this calculation.

Target StdDev is the standard deviation of the target when this state has the highest membership probability. Cases in which there is a tie for maximum (almost impossible for continuous data) do not enter into this calculation.

State	Percent	Correlation	Target mean	Target StdDev
1	23.76	-0.53350	-0.54538	0.45473
2	21.73	-0.52368	-0.71809	0.56342
3	34.03	0.38210	0.35674	0.47747
4	20.48	0.62840	0.92173	0.49069

The reader should look back at the table of RAND1 and RAND2 means for each of the four states and confirm that the correlations and target means shown in the preceding table make sense. We also see that the state membership probabilities conform with the transition matrix. As expected for random series, the target standard deviations are all about the same.

Last but not least is the list of models, sorted in descending order of their multiple-R with the target. As expected (or at least hoped), the models involving either RAND1 or RAND2 appear first, and they are all extremely significant. As soon as these two variables are exhausted, multiple-R plunges and significance is lost. The remainder of this table is not shown here, but this situation continues.

------> Hidden Markov Models correlating with SUM12 <------

Predictor 1	Predictor 2	Multiple-R	Solo pval	Unbiased
RAND1	RAND2	0.8896	0.0010	0.0010
RAND1	RAND3	0.6937	0.0010	0.0010
RAND1	RAND5	0.6680	0.0010	0.0010
RAND0	RAND1	0.6619	0.0010	0.0010
RAND1	RAND9	0.6604	0.0010	0.0010
RAND1	RAND8	0.6590	0.0010	0.0010
RAND2	RAND5	0.6579	0.0010	0.0010
RAND0	RAND2	0.6554	0.0010	0.0010
RAND2	RAND9	0.6493	0.0010	0.0010
RAND1	RAND7	0.5870	0.0010	0.0010
RAND1	RAND4	0.5845	0.0010	0.0010
RAND2	RAND4	0.5756	0.0010	0.0010
RAND2	RAND3	0.5721	0.0010	0.0010
RAND2	RAND7	0.5667	0.0010	0.0010
RAND2	RAND6	0.5648	0.0010	0.0010
RAND2	RAND8	0.5623	0.0010	0.0010
RAND1	RAND6	0.3938	0.0010	0.0010
RAND3	RAND9	0.0307	0.1110	0.8760

A Sensible and Practical Example

This section demonstrates an example of hidden Markov models using actual data, in this case an application that predicts future movement of a financial market. There are five candidates for predictor variables and a single target:

CMMA_5 is the current closing price of the market, minus its 5-day moving average. This shows the degree to which the market just (as of the end of the current day) departed from its recent price level.

CMMA_10 is a similar quantity, but based on the 10-day moving average.

CMMA_20 is a similar quantity, but based on the 20-day moving average.

LIN_ATR_7 is the slope of the best-fit straight line connecting the prices over the most recent 7 days, normalized by average true range. This indicates the short-term price trend in the market.

LIN_ATR_15 is a similar quantity, but based on the 15-day trend.

DAY_RETURN_1 is the market change over the next day, normalized by average true range. This variable serves as the target, as it represents the future change of the market price.

This example specifies that two predictors will be used by the model and three states are possible. The model that correlates most highly with the target uses CMMA_5 and CMMA_20 as predictors. The means and standard deviations of these variables are shown for each of the three states:

Means (top number) and standard deviations (bottom number)

State	CMMA_20	CMMA_5
1	-20.81845	-15.87819
	9.42582	16.57821
2	24.57826	17.83951
	8.25328	15.22672
3	3.57633	2.36846
	7.27092	17.76842

The three states are highly distinct in terms of their predictor distributions. CMMA_20, in particular, has means that are widely separated relative to their standard deviations. We see that State 1 is characterized by today's price being much lower than recent prices, State 2 is characterized by today's price being much higher than recent prices, and State 3 is characterized by today's price being about the same as recent prices. This sounds almost too "sensible" to be believed, but numerous reruns of the test consistently produced similar results.

The transition probability matrix, shown in the following, reveals several interesting properties. First, we see that states have considerable persistence; there is about a 90 percent probability that tomorrow the process will remain in the same state as today. What is also interesting is that it is nearly impossible for the market to transition between States 1 and 2 without going through State 3, and in fact probably staying in State 3 for some time. In fact, the probability of going from State 1 to State 2 is zero to at least four digits!

Transition probabilities...

	1	2	3
1	0.8978	0.0000	0.1022
2	0.0014	0.9095	0.0890
3	0.0711	0.0747	0.8542

The table of additional properties shows how these states relate to the target, the price change of the market the next day. We see that State 3, that corresponding to prices remaining fairly constant, is the most common, occurring almost 40 percent of the time. We also see at least 1-day persistence of price movements into the future, as State 1, which corresponds to a pattern of today's closing price being far below recent prices, is associated with a negative price movement tomorrow. Similarly, State 2, which corresponds to a pattern of today's closing price being far above recent prices, is associated with an upward price movement tomorrow. Finally, it is noteworthy that the standard deviation of the target when in State 1 is almost 50 percent higher than when in the other two states. Thus, we can expect unusually large market turbulence when we have been in a pattern of prices closing far below their recent values. This agrees well with intuition, but it is nice to see it corroborated numerically.

State	Percent	Correlation	Target mean	Target StdDev
1	27.75	-0.07034	-0.05099	0.86047
2	32.41	0.06831	0.08906	0.60901
3	39.84	-0.00049	0.02438	0.64007

Finally, we have the list of models sorted according to their relationship with the target. The major takeaway from this list is that the CMMA variables are much more important to predicting tomorrow's price movement than the linear trend variables. Also, the degree of significance of these relationships is impressive, usually the minimum obtainable from the 1000 Monte-Carlo replications performed.

Predictor 1	Predictor 2	Multiple-R	Solo pval	Unbiased
CMMA_20	CMMA_5	0.0807	0.0010	0.0010
CMMA_5	LIN_ATR_7	0.0762	0.0010	0.0010
CMMA_10	CMMA_5	0.0689	0.0010	0.0010
CMMA_10	CMMA_20	0.0686	0.0010	0.0010
CMMA_20	LIN_ATR_7	0.0650	0.0010	0.0010
CMMA_20	LIN_ATR_15	0.0442	0.0010	0.0010
CMMA_10	LIN_ATR_7	0.0408	0.0010	0.0010
CMMA_10	LIN_ATR_15	0.0330	0.0020	0.0080
CMMA_5	LIN_ATR_15	0.0227	0.0480	0.1500
LIN_ATR_15	LIN_ATR_7	0.0168	0.1790	0.4750

CHAPTER 5

Stepwise Selection on Steroids

It's likely that everyone reading this book is familiar with stepwise selection. Typically, you have a large set of candidates for some task, often prediction or classification. You test each individual candidate and select the one candidate that performs the task best. Then you test the remaining candidates, seeking the one that, in conjunction with the one already selected, performs best. This is repeated as desired. It is a fast, efficient, and usually fairly effective way of selecting a respectable subset of features from a potentially large population.

Unfortunately, this venerable and widely used algorithm suffers from several serious weaknesses. The most obvious and problematic is that very often an application can be handled only when we have multiple features available simultaneously. As a crude example, suppose we wish to evaluate the intelligence of a person. We could give this person a test involving sophisticated logical reasoning. Suppose the person got half of the problems correct. That score would mean one thing if the person were 25 years old, and it would mean something else entirely if the person were 3 years old. Or suppose we want to measure a risk of cardiac disease. Neither height alone nor weight alone would be very good, but the two together would provide significant predictive power. When we are dealing with such an application, simple stepwise selection could easily miss a predictor that is immensely powerful when used in conjunction with another predictor but that is nearly worthless when used alone.

Another issue with stepwise selection that can be a problem if not properly handled is the fact that a naive selection criterion results in the performance steadily increasing as we add more variables (features). This is due to the fact that random noise is mistaken for legitimate information. The selection algorithm gets better and better at learning the properties of the noise as more features are examined, all the while blissfully unaware

© Timothy Masters 2020
T. Masters, *Modern Data Mining Algorithms in C++ and CUDA C*,
https://doi.org/10.1007/978-1-4842-5988-7_5

that the supposedly valuable "features" do not represent repeatable patterns. If we judge quality on a simplistic measure such as in-sample performance, we are very likely to add more variables than are appropriate and actually decrease out-of-sample performance.

Yet another potential problem with naive stepwise selection is failure to distinguish between seemingly good performance versus statistically sound performance. An apparently great performance figure means nothing if there is a substantial probability that it could have done that well by nothing more than good luck. These are the key issues that will be addressed in this chapter.

In particular, this chapter will present a generic, broadly applicable stepwise selection algorithm, illustrated with complete source code. This code will use a quickly trainable nonlinear model to evaluate the predictive power of feature sets. However, it will be provided in such a way that the reader can easily drop in an alternative model, or even wrap this algorithm around the reader's existing modeling software. The three aspects of this stepwise selection algorithm that set it apart from simple, traditional methods are as follows:

- It significantly overcomes the problem of missing important variables that have little value when used alone, while avoiding the combinatoric explosion arising from exhaustive testing of all possible subsets. It does this by saving multiple promising subsets at every step and evaluating future candidates in conjunction with these multiple subsets.

- It avoids the "more variables means better performance" issue by judging the quality of a feature set according to its cross-validation performance. This tremendously reduces the likelihood that random noise will be misinterpreted as valid predictive information. It also provides a simple and effective automatic way to stop adding new features to the feature set.

- As each new feature variable is added, it computes two probabilities. The most important is the probability that if all currently selected features are truly worthless, the performance criterion achieved by the current feature set could be as good as it is by pure luck. A less important but still useful measure is the probability that if all current features are truly worthless, the performance *increase* provided by adding the most recently selected feature to those already selected could have been as large as we observed.

Complete source code for the algorithms discussed in this chapter is in
STEPWISE.CPP.

The Feature Evaluation Model

In order to present the enhanced stepwise selection algorithm with actual working
source code, we need a basis model with which to assess the predictive power of
feature variables. It need not be especially powerful or sophisticated, because its role is
secondary to our main goal: stepwise feature selection. But it should be powerful enough
to allow a fair demonstration of the algorithm. And of course it should enjoy fast enough
training so as to allow practical testing of the algorithm.

One of my favorite prediction algorithms fits the bill nicely. This is what is sometimes
called linear-quadratic regression, or perhaps quadratic-linear regression. In this model,
the input vector is expanded to include not just the feature variables, but their squares
and all possible cross products. These variables are supplied to an ordinary linear
regression model. This hybrid approach gives us the speed and stability of simple linear
regression while still supplying significant nonlinear capabilities, including complete
reversal of the predictor/target relationship across the feature domain, as well as
complex interactions between features. It's really a wonderful model.

Mathematically, standardization of the input variables is not required and makes no
theoretical difference in performance. However, for real-life computing, as well as easy
human interpretation of model coefficients, it is important to standardize the inputs so
that their means are all zero and their variances are equal (one in this code).

A conscientious user will ensure that their input features are at least somewhat
independent. Significant correlation among features is unavoidable and perfectly
acceptable, but exact or nearly exact collinearity is verboten. Nevertheless, humans are
fallible. For this reason, I do not employ the textbook matrix inversion method of linear
regression. Instead, I use singular value decomposition. This is not the forum
for a detailed discussion of this technique; it is widely discussed elsewhere. The file
SVDCMP.CPP contains the class for doing this, and the comments at the beginning of
the file explain in detail how to use it for "safe" linear regression.

Code for the Foundation Model

The stepwise selection code of this chapter is designed in such a way that it is easy for the reader to drop in an alternative model if desired. Two routines are required, one to train the model given a training set and another to compute the model's performance criterion given a test set. As far as the stepwise algorithm shown here is concerned, these two routines are black boxes; what goes on inside them is irrelevant. Thus, the reader can replace my model with whatever model they wish. In fact, alternatives are not limited to prediction models. An alternative can be any process that optimizes some measure of some sort of performance on a training set and then evaluates that same performance measure on a test set. The only restriction is that the performance measure must be a "total" in the sense that the performance (or error) of a set of cases is the sum of the performance for the individual cases in the set.

The calling scheme for the training and testing routines are as follows:

```
static int fit_model (
    int ncases ,        // Number of cases in data
    int omit_start ,    // Index of first case to omit from fitting
    int omit_stop ,     // One past last to omit
    int npred ,         // Number of predictors to use
    int *preds ,        // Their indices in data
    int ncols ,         // Number of columns (variables) in data
    double *data ,      // ncases by ncols dataset
    double *target ,    // ncases vector of targets
    double *coefs       // npred + npred * (npred+1) / 2 + 1 computed coefs
    )

static double evaluate (
    int ncases ,        // Number of cases in data
    int test_start ,    // Index of first case to test
    int test_stop ,     // One past last to test
    int npred ,         // Number of predictors to use
    int *preds ,        // Their indices in data
    int ncols ,         // Number of columns (variables) in data
    double *data ,      // ncases by ncols dataset
```

```
double *target , // ncases vector of targets
double *coefs   // npred + npred * (npred+1) / 2 + 1 input coefs
)
```

The dataset data contains ncases rows of observations, with each row containing ncols variables, not all of which need to be used. In fact, the routines will use npred of these variables, with the entries in preds designating the columns (origin zero) used. A target variable target would be the true value of the quantity being predicted when this is a predictive model (as it is here), but in a more general scenario, it can be any quantity that the training and evaluation routines can use for optimization and quality measurement.

For the training routine fit_model(), we specify the starting row omit_start as well as omit_stop, which is one greater than the stopping row for a block of cases to be ignored during training. The evaluation routine similarly is given a starting row (test_start) as well as one past the stopping row (test_stop) for the cases that will be used to evaluate the quality of the model.

Finally, we must provide an array that will contain the optimized specifications of the model or whatever is needed to define the scheme being implemented. Here, this array contains the coefficients of the linear-quadratic model. This array will be output from fit_model() and serve as input to evaluate(). Here is the complete code for fit_model(). Note our need for three static variables that frequently enable reuse of the routine without having to delete the SingularValueDecomp object and then allocate a new one. Creation and destruction of this object requires multiple memory allocations/deallocations and hence should be avoided whenever possible to avoid memory thrashing.

```
static int svdcmp_nrows = 0 ;   // These three statics let us often preserve the
static int svdcmp_ncols = 0 ;    // SingularValueDecomp object for re-use
static SingularValueDecomp *sptr = NULL ;

static int fit_model ( ... )   // The calling list was shown above
{
  int icase, ivar, nvars, ntrain, k1, k2 ;
  double *aptr, *bptr, *dptr ;

  nvars = npred + npred * (npred+1) / 2 ;      // Linear + quadratic
  ntrain = ncases - (omit_stop - omit_start) ; // Number of training cases
  if (sptr == NULL || svdcmp_nrows != ntrain || svdcmp_ncols != nvars+1) {
    if (sptr != NULL)   // We only recreate the object when its dimensions change
      delete sptr ;
```

```
    sptr = new SingularValueDecomp ( ntrain , nvars+1 , 0 ) ;
    svdcmp_nrows = ntrain ;
    svdcmp_ncols = nvars + 1 ;
    }

  if (sptr == NULL || ! sptr->ok)
    // Handle insufficient-memory failure here; see STEPWISE.CPP for how I do it
```

We now build the design matrix in the a member. We'll soon put the target in the b member. It would be slightly more efficient to make that a part of this same loop, but for clarity, I do it separately, later.

```
  aptr = sptr->a ;
  bptr = sptr->b ;

  for (icase=0 ; icase<ncases ; icase++) {
    if (icase >= omit_start && icase < omit_stop)
      continue ;   // Skip the block that will be tested after this training

    dptr = data + icase * ncols ;
    k1 = 0 ;  // These will index the cross products
    k2 = 1 ;
    for (ivar=0 ; ivar<nvars ; ivar++) {
      if (ivar < npred)                  // Linear terms
        *aptr++ = dptr[preds[ivar]] ;
      else if (ivar < 2*npred)           // Square terms
        *aptr++ = dptr[preds[ivar-npred]] * dptr[preds[ivar-npred]] ;
      else {                             // Cross-product terms
        *aptr++ = dptr[preds[k1]] * dptr[preds[k2]] ;
        ++k2 ;                           // Advance to the next cross product term
        if (k2 == npred) {
          ++k1 ;
          k2 = k1 + 1 ;
          }
        }
      }
    *aptr++ = 1.0 ; // Constant
    }
```

174

The design matrix is built. Compute its singular value decomposition. Copy the target to the b member and then perform the back-substitution to compute the coefficients. As stated earlier, it would have been slightly more efficient to copy the target to b in the same loop in which we built the design matrix. But that loop was complicated enough already, so to avoid adding even more complexity, I do it separately, here. Feel free to move it up to that loop.

The constant of 1.e-7 is not terribly critical, coming into play only if a careless user supplied predictor variables that have very high collinearity. This constant serves as a threshold for dealing with such a situation. For more information on what's happening in regard to the SingularValueDecomp object, see SVDCMP.CPP. An in-depth treatment of the subject can easily be found online and in many textbooks.

```
sptr->svdcmp () ;

for (icase=0 ; icase<ncases ; icase++) {
  if (icase < omit_start || icase >= omit_stop)
    *bptr++ = target[icase] ;
  }

sptr->backsub ( 1.e-7 , coefs ) ;
return 0 ;
}
```

Code for evaluating the performance of a trained model is straightforward. We will sum the mean squared error across all test cases. If you change the foundation model or performance criterion, you must make sure that your performance measure is "summable" like this, meaning that the performance for a set of cases is the sum of the performance of individual cases. If this is not possible, you will need to modify the cross-validation routine to be described in the next section. Here is the evaluation code:

```
static double evaluate ( ... )    // The calling list was shown earlier
{
  int icase, ivar, nvars, k1, k2 ;
  double x, *dptr, pred, diff, err ;

  nvars = npred + npred * (npred+1) / 2 ; // Linear + quadratic
```

The following loop uses the trained model to compute the predicted value for each case in the test set. The true value is subtracted from this predicted value, the difference is squared, and this squared error is summed across the test set. This sum is returned as the performance criterion.

```
err = 0.0 ;   // Will sum the squared error

for (icase=test_start ; icase<test_stop ; icase++) {   // Pass through the test set
  dptr = data + icase * ncols ;                        // This test case

  k1 = 0 ;    // These will index the cross products
  k2 = 1 ;

  pred = coefs[nvars] ;        // The constant term in the prediction equation

  for (ivar=0 ; ivar<nvars ; ivar++) {

    if (ivar < npred)              // Linear terms
      x = dptr[preds[ivar]] ;

    else if (ivar < 2*npred)     // Square terms
      x = dptr[preds[ivar-npred]] * dptr[preds[ivar-npred]] ;

    else {                       // Cross-product terms
      x = dptr[preds[k1]] * dptr[preds[k2]] ;
      ++k2 ;                     // Advance to the next cross product term
      if (k2 == npred) {
        ++k1 ;
        k2 = k1 + 1 ;
        }
      }

    pred += x * coefs[ivar] ;    // Sum the prediction equation
    } // For ivar

  diff = pred - target[icase] ;  // Predicted minus true
  err += diff * diff ;           // Sum the squared error
  } // For icase

return err ;
}
```

The Cross-Validated Performance Measure

The naive and traditional way of selecting features for a task is to maximize an *in-sample* performance criterion. In other words, we use a single dataset to compute the performance criterion and select those features that provide the most optimal criterion. Of course, an even modestly responsible developer will then go on to use a second, independent data sample to evaluate the quality of that feature set in conjunction with the model that was employed. But by then it's too late. That feature selection method will almost always produce a suboptimal feature set.

The reason this naive selection method is suboptimal is that any dataset is a mixture of legitimate information and random noise. Unfortunately, in virtually any application, there is no way for the optimization algorithm to distinguish between noise and legitimate information when it has only one dataset to examine. Thus, whatever algorithm associates features in the dataset with correct target values in order to compute a performance measure will, to at least some degree, confuse noise with features. By definition, noise will not repeat in other data, and so some features will be selected based on their ability to relate noise to the target, a dangerous error.

Many ways to deal with this serious issue have been devised. Most are based on some sort of complexity penalty. The performance criterion may be based on something simple like applying a penalty that increases as more features are added. Others may try to evaluate the degree to which features contribute to performance within the dataset and reject those features that appear to make relatively little contribution. Still others may use sophisticated measures of complexity and penalize feature sets that produce a model with high complexity. These are all worthy endeavors, but they all *indirectly* address the issue of feature selection confusing nonrepeatable noise with repeatable information.

In my opinion, we are best off taking a direct approach to solving this problem: use one dataset to optimize our core model's performance with a trial feature set and then evaluate the quality of this feature set by measuring its performance on a *different* dataset. This way, features that capture legitimate information will also perform well on the second dataset, in which the legitimate information also appears. But features that mistake random noise for legitimate information will perform poorly on the second dataset, because those phony patterns will likely not appear.

We would waste a lot of potentially expensive data if we simply divided our dataset in half for this purpose. So instead we use cross validation. A fraction of the dataset is withheld from optimizing the model, and that withheld portion is used to test the trained

177

model. Then that portion is returned to the dataset and a different fraction is withheld. This repeats in such a way that each case in the dataset appears in a withheld portion exactly once.

One unavoidable disadvantage of cross validation is that it requires a sometimes annoying tradeoff. If we hold out only a few cases at a time (with each in/out split being called a *fold*), processing time will be huge, because we have to reoptimize the model for each fold. Thus, we are encouraged to minimize the number of folds (hold out many cases each time). But if we hold out a lot of cases, we reduce the number of cases used for optimization, which makes the model less accurate and less stable, leading to less accurate results. The rule of thumb is that we should use as many folds as possible, consistent with being able to run the program in a manageable length of time.

If you are working with canned or black-box training/testing routines, cross validation usually requires massive shuffling of the dataset to keep training and test sets contiguous. However, because we are in control of these procedures in this example, we can completely avoid shuffling. During training we specify the starting and stopping locations of the withheld portion of the dataset (omit_start and omit_stop in the code seen earlier), and during testing we specify these same locations in the test routine (test_start and test_stop in the prior code). The work of the cross-validation routine is thereby simplified.

The cross-validation routine by which we evaluate the quality of a feature subset is shown here, with a brief discussion. All of the calling parameters have been seen before.

```
static int xval (
   int ncases ,      // Number of cases in data
   int nfolds ,      // Number of folds
   int npred ,       // Number of predictors to use
   int *preds ,      // Their indices in data
   int ncols ,       // Number of columns (variables) in data
   double *data ,    // ncases by ncols dataset
   double *target ,  // ncases vector of targets
   double *coefs ,   // npred + npred * (npred+1) / 2 + 1 work area
   double *crit      // Computed criterion returned here
   )
{
   int ifold, n_remaining, test_start, test_stop ;
   double error ;
```

```
  n_remaining = ncases ;
  test_start = 0 ;

  error = 0.0 ;
  for (ifold=0 ; ifold<nfolds ; ifold++) {
    test_stop = test_start + n_remaining / (nfolds - ifold) ;
    fit_model( ncases, test_start, test_stop, npred, preds, ncols, data, target, coefs ) ;
    error += evaluate ( ncases , test_start , test_stop , npred , preds , ncols , data ,
                        target , coefs ) ;
    n_remaining -= test_stop - test_start ;
    test_start = test_stop ;
    }

  *crit = 1.0 - error / ncases ;
  return 0 ;
}
```

This routine keeps in n_remaining the number of cases remaining to be tested. For each fold, the number of cases tested is the number remaining to be tested, divided by the number of folds remaining (nfolds - ifold). We add this quantity to the starting index of the test block to get (one past) the ending index. Those cases outside the test block are used to optimize the model, which is then evaluated using the test block. The next test block starts just past the prior one. Because the targets have been standardized to unit variance, it's easy to compute and use R-square as the performance criterion, as shown.

The Stepwise Algorithm

The algorithm that preserves multiple "best so far" feature sets is trickier than it might seem. I will now present this algorithm as both simplified pseudo-code and actual C++ code. The overall flow looks like this:

```
Allocate all arrays
Save a copy of the targets
While adding variables...
   For all permutations [irep from 0 through mcpt_reps-1]...
      Fetch targets from saved copy
      if (irep not zero) [If this is a permutation trial]
         Set random seed for this permutation
         Shuffle
      Get a pointer to private 'so far' variables for this irep
      Add a variable, updating private 'so far' variables
      If crit decreased AND we have min vars AND not permutation, quit
      If irep=0 [This is the unpermuted trial]
         Initialize for MCPT
      Else
         Update MCPT
   Print variable just added, along with its MCPT p-values
   if we reached user's max vars, break
Free all arrays
```

Complete source code for this algorithm as well as the routines already shown is in the file STEPWISE.CPP. I'll break it down here into individual blocks of code, discussing each as it is presented.

In the next listing we see the calling parameters for the coordination routine step_main(). Most of the parameters are self-explanatory, but a few need special mention. The user parameters nkept and nfolds are the number of "best so far" feature sets retained at each step and the number of cross-validation folds, respectively. If nkept is set to 1, we have ordinary forward selection (though still based on the cross-validation criterion rather than in-sample performance).

The user parameters minpred and maxpred specify the minimum number of features demanded by the user and the maximum, respectively. We will keep at least minpred features, even if getting this many results in the performance criterion deteriorating. I recommend setting this to 1 for most applications and setting maxpred to the number of candidates, being confident that in most situations addition of features will halt long before this limit is reached. Finally, know that pred_indices has nothing whatsoever to do with the algorithm. I included it only so that if the candidates have been extracted from a larger "master" database, and the names of these master database variables are available, we can print these names for the user. You will see how this works later, and you can remove or modify that code if you so wish. Here is the calling convention:

```
static int step_main (
  int ncases ,            // Number of cases in data
  int ncand ,             // Number of predictor candidates
  int *pred_indices ,     // Their indices in original database (for naming only)
  double *predvars ,      // ncases by ncand matrix of predictor candidates
  double *target ,        // ncases vector of targets
  int nkept ,             // Number of best predictors retained in each step
  int nfolds ,            // Number of folds for XVAL criterion
  int minpred ,           // Minimum number of predictors in final model
  int maxpred ,           // Maximum number of predictors in final model
  int mcpt_type ,         // 1=complete, 2=cyclic
  int mcpt_reps ,         // Number of MCPT replications, <=1 for no MCPT
  int *n_in_model ,       // Returns final number of predictors in model
  int *model_vars         // Returns indices in predvars of model variables
  )
```

The first step is to allocate all arrays that we will need. In order to avoid unnecessary retraining, we will keep in already_tried a record of all models trained at each step, every retained set (nkept) with every new candidate (ncand), each up to maxpred long. We need tried_length ints for that.

```
tried_length = maxpred * ncand * nkept ;
already_tried = (int *) MALLOC ( tried_length * sizeof(int) ) ;
trial_vars = (int *) MALLOC ( maxpred * sizeof(int) ) ;
all_best_trials = (int *) MALLOC ( mcpt_reps * maxpred * nkept * sizeof(int) ) ;
prior_best_trials = (int *) MALLOC ( maxpred * nkept * sizeof(int) ) ;
all_best_crits = (double *) MALLOC ( mcpt_reps * nkept * sizeof(double) ) ;
prior_best_crits = (double *) MALLOC ( nkept * sizeof(double) ) ;
coefs = (double *) MALLOC ( (maxpred + maxpred * (maxpred+1) / 2 + 1) *
                            sizeof(double) ) ;
target_copy = (double *) MALLOC ( 2 * ncases * sizeof(double) ) ;
```

We save a copy of the targets so we can get them back after permuting, and then we have two nested loops. The outer loop adds variables one at a time, and the inner loop handles Monte-Carlo permutations as each new variable is added. The first step in the permutation loop is to fetch the original, unpermuted targets. If we are in a permutation pass (all but the first), set the random seed according to the replication number. This ensures that as we add variables, each replication is working with exactly the same permutation of the targets. Count included features in n_so_far.

```
memcpy ( target_copy , target , ncases * sizeof(double) ) ;
n_so_far = 0 ; // Counts number of variables so far

for (;;) {        // This loop adds variables, one with each pass through the loop
  for (irep=0 ; irep<mcpt_reps ; irep++) {   // Permutation loop
    memcpy ( target_copy , target , ncases * sizeof(double) ) ; // Get original target

    if (irep) {                     // If doing permuted runs, shuffle
      irand = 17 * irep + 11 ;      // Always use the same shuffle in each rep
      fast_unif ( &irand ) ;        // Warm up the generator
      fast_unif ( &irand ) ;        // Warm up the generator again
      if (mcpt_type == 1) {         // Complete permutation (best, for independent targets)
        i = ncases ;                // Number remaining to be shuffled
        while (i > 1) {             // While at least 2 left to shuffle
          j = (int) (fast_unif ( &irand ) * i) ;
          if (j >= i)
            j = i - 1 ;
          dtemp = target_copy[--i] ;
          target_copy[i] = target_copy[j] ;
          target_copy[j] = dtemp ;
          }
        } // Type 1, Complete
      else if (mcpt_type == 2) { // Cyclic; Inferior but mandatory if serial correlation
        j = (int) (fast_unif ( &irand ) * ncases) ;
        if (j >= ncases)
          j = ncases - 1 ;
        for (i=0 ; i<ncases ; i++)
          target_work[i] = target_copy[(i+j)%ncases] ;
        for (i=0 ; i<ncases ; i++)
          target_copy[i] = target_work[i] ;
        } // Type 2, Cyclic
      } // If in permutation run (irep > 0)
```

We are now ready to add a variable. Each of the mcpt_reps permutation replications must be internally consistent and independent. We already set the random permutation seed according to the replication number, ensuring that each replication will have a consistent reordering of the targets. The variable addition routine will also have two arrays (best_trials and best_crits) that must be maintained as variables are added, so we need a separate work area for these arrays for each permutation replication.

```
best_trials = all_best_trials + irep * maxpred * nkept ; // Private for this replication
best_crits = all_best_crits + irep * nkept ;

if (n_so_far == 0)
   prior_crit = -1.e60 ;        // Make sure the first variable is kept
else
   prior_crit = best_crits[0] ;

n_this_rep = n_so_far ;     // This will be incremented by add_var()

ret_val = add_var ( ncases , ncand , predvars , target_copy , nkept , nfolds ,
                    mcpt_reps , &n_this_rep , best_trials , best_crits ,
                    prior_best_trials , prior_best_crits , trial_vars ,
                    already_tried , coefs ) ;
```

The criterion achieved by adding this latest variable is in best_crits[0], as will become clear later. If adding this variable caused the performance criterion to decrease, and if prior to adding this new variable we had at least as many as the user's specified minimum, and if we are in the first, unpermuted replication, we reject this new variable and stop adding new variables. We're done. Fetch the best models prior to adding this variable, even though we probably will need only the best of these.

```
if (best_crits[0] <= prior_crit && n_so_far >= minpred && ! irep) {
   for (i=0 ; i<nkept ; i++) {
      best_crits[i] = prior_best_crits[i] ;
      memcpy ( &best_trials[i*n_so_far] ,
               &prior_best_trials[i*n_so_far] ,
               n_so_far * sizeof(int) ) ;
   }
   // Here you should tell the user that we terminated due to deterioration
   goto STEP_MAIN_FINISH ;
}
```

The last step in the permutation loop is to handle things related to the two Monte-Carlo permutation tests. Normally, the R-square criterion will never be negative. But in unusual situations in which the model does *worse* than guessing the prediction to be the target mean (usually the best possible guess), the criterion can be negative. For the permutation test, we limit the criterion to zero, as this produces a more conservative test by encouraging ties.

If we have just done the initial, unpermuted replication, we save the original test criteria, which are the performance criterion and the change in the criterion. But if we are now in a permuted replication, we update the counters.

```
if (prior_crit < 0.0)
   prior_crit = 0.0 ;
new_crit = best_crits[0] ;
if (new_crit < 0.0)
   new_crit = 0.0 ;

if (irep == 0) {
   original_crit = new_crit ;
   original_change = new_crit - prior_crit ;
   mcpt_mod_count = mcpt_change_count = 1 ;
   }

else {
   if (new_crit >= original_crit)
      ++mcpt_mod_count ;
   if (new_crit - prior_crit >= original_change)
      ++mcpt_change_count ;
   }

} // for irep (MCPT loop)
```

At this point, we have added a single variable and performed all MCPT replications for this new variable. This lets us test two null hypotheses, both of which have the same assumption that all candidates are truly worthless. One test computes the probability that our model, which includes all variables selected so far, could have obtained performance at least as good as what we observed by luck alone. The other computes the probability that, by luck alone, adding this latest variable could have produced as large a performance improvement as we observed.

Both of these computations assume that one or both of two conditions is true. Either *all* candidates must be free of serial or any other correlation (independent) or the target variable's observations must be independent. If this requirement is not satisfied, the computed probabilities will be too small, a very dangerous situation because it inspires false confidence in our results. In this situation, using the otherwise inferior *cyclic* permutation instead of the superior *complete* permutation can somewhat alleviate this problem, but it can never be completely eliminated.

Here is the code for computing and printing results for this most recently added variable:

```
if (n_so_far == 0)   // For the first variable, these are identical tests
   mcpt_change_count = mcpt_mod_count ;

sprintf_s ( msg , "%8.4lf %8.3lf %9.3lf ", original_crit,        // R-square so far
         (double) mcpt_mod_count / (double) mcpt_reps,           // Model p-value
         (double) mcpt_change_count / (double) mcpt_reps ) ;  // Change p-value

for (i=0 ; i<n_this_rep ; i++) {  // Note how I use pred_indices to get variable names
   sprintf_s ( msg2 , " %s", var_names[pred_indices[all_best_trials[i]]] ) ;
   strcat_s ( msg , msg2 ) ;      // Feel free to change this naming code as needed
   }
audit ( msg ) ;

++n_so_far ;                      // We just added one variable
if (n_so_far == maxpred)          // Are we all done adding variables?
   break ;
} // Endless for loop adding variables

STEP_MAIN_FINISH:                 // We get here either by breaking out of var loop or error
 *n_in_model = n_so_far ;         // Return optimal predictors to caller
 for (i=0 ; i<n_so_far ; i++)
   model_vars[i] = all_best_trials[i] ;

// Free all allocated arrays here
   return ret_val ; // This contains a return code, normally zero, nonzero for error
}
```

Finding the First Variable

In the code just seen, we call add_var() to add a single variable to the model so far. This is just a small wrapper routine that splits the task into finding the first variable versus adding a new variable to an existing model:

```
static int add_var (
   int ncases ,            // Number of cases in data
   int ncand ,             // Number of predictor candidates
   double *predvars ,      // ncases by ncand matrix of predictor candidates
   double *target ,        // ncases vector of targets
   int nkept ,             // Number of best predictors retained in each step
   int nfolds ,            // Number of folds for XVAL criterion
   int mcpt_reps ,         // Purely for progress updates
   // Items below are maintained throughout entire stepwise process
   int *n_so_far ,         // Number of predictors so far
   int *best_trials ,      // Variable indices of the nkept best models at this step
   double *best_crits ,    // Performance criteria of the nkept best models at this step
   // Items below are strictly work areas, not preserved
   int *prior_best_trials ,
   double *prior_best_crits ,
   int *trial_vars ,
   int *already_tried ,
   double *coefs
   )
{
   int ret_val ;

   if (*n_so_far == 0) {
      *n_so_far = 1 ;
      ret_val = first_var ( ncases , ncand , predvars , target , nkept , nfolds , mcpt_reps ,
                     best_trials , best_crits , trial_vars , coefs ) ;
      }
   else
      ret_val = next_var ( ncases , ncand , predvars , target , nkept , nfolds , mcpt_reps ,
```

```
            n_so_far , best_trials , best_crits , prior_best_trials , prior_best_crits ,
            trial_vars , already_tried , coefs ) ;
   return ret_val ;
}
```

We now discuss first_var(), which finds the nkept best single variables, thereby giving us the best single variable as well as the set of promising close competitors to pass on when we add the second variable. Here is the calling parameter list:

```
static int first_var (
   int ncases ,          // Number of cases in data
   int ncand ,           // Number of predictor candidates
   double *predvars ,    // ncases by ncand matrix of predictor candidates
   double *target ,      // ncases vector of targets
   int nkept ,           // Number of best predictors retained in each step
   int nfolds ,          // Number of folds for XVAL criterion
   int mcpt_reps ,       // Purely for progress updates
   // Items below are maintained throughout entire stepwise process
   int *best_trials ,    // Variable indices of the nkept best models at this step
   double *best_crits ,  // Performance criteria of the nkept best models at this step
   // Items below are strictly work areas, not preserved
   int *trial_vars ,
   double *coefs
   )
```

We will be saving the best nkept models. Initialize the array of them to be empty. Since this is the first variable, each model will be just one variable.

```
{
   int i, j, this_var ;
   double crit ;

   for (i=0 ; i<nkept ; i++) {
      best_crits[i] = -1.e60 ;
      best_trials[i] = -1 ;   // Not needed for algorithm but eliminates 'use-before-set' error
      }
```

This main loop evaluates the performance of every candidate feature variable. The criterion routine xval() was described on Page 177.

```
for (this_var=0 ; this_var<ncand ; this_var++) {

  trial_vars[0] = this_var ;      // There is just this one variable in this model
  if (xval ( ncases , nfolds , 1 , trial_vars , ncand , predvars , target , coefs , &crit ))
    return ERROR_INSUFFICIENT_MEMORY ;
```

We now have in crit the R-square performance criterion for trial variable this_var, which is the index of the column in predvars. In ordinary stepwise selection, this is all we need. But in this advanced method, we need to keep the nkept best competitors so that when we go on to add a second variable, we can test each new candidate in combination with each of these best first variables.

The easiest way to maintain this array is to keep them in descending order of performance. We do this in three steps. First, pass through the rankings that we have so far and see where this one fits in. Second, if it beats any that we are already keeping, we move down the ones that are kept so far but that are inferior to this one, freeing up an empty slot. Last, we put this variable in the empty slot.

Note that the best_trials and best_crits arrays will be maintained throughout the entire stepwise process. Thus, when we do the permutation runs, in order to maintain continuity, each permutation must have its own private copy of each of these two arrays. They cannot be shared among the permutations.

```
  // Perhaps insert this in array of best

  for (i=0 ; i<nkept ; i++) { // See where it fits in ranking
    if (crit > best_crits[i])
      break ; // It beats the one in slot i
    }

  if (i < nkept) { // We need to move down and insert
    for (j=nkept-2 ; j>=i ; j--) {
      best_trials[j+1] = best_trials[j] ;
      best_crits[j+1] = best_crits[j] ;
      }
```

```
        best_trials[i] = this_var ; // Insert this one
        best_crits[i] = crit ;
        } // If we are inserting this new 'best' model

    } // For this_var, trying all competitors

  return 0 ;
}
```

Adding a Variable to an Existing Model

This is the most complex part of the entire stepwise procedure, because a lot of bookkeeping is involved. We'll take it step by step. Here is its calling parameter set:

```
static int next_var (
  int ncases ,           // Number of cases in data
  int ncand ,            // Number of predictor candidates
  double *predvars ,     // ncases by ncand matrix of predictor candidates
  double *target ,       // ncases vector of targets
  int nkept ,            // Number of best predictors retained in each step
  int nfolds ,           // Number of folds for XVAL criterion
  int mcpt_reps ,        // Purely for progress updates
  // Items below are maintained throughout entire stepwise process
  int *n_so_far ,        // Number of predictors so far; out=in+1
  int *best_trials ,     // Variable indices of the nkept best models at this step
  double *best_crits ,   // Performance criteria of the nkept best models at this step
  // Items below are strictly work areas, not preserved
  int *prior_best_trials ,
  double *prior_best_crits ,
  int *trial_vars ,
  int *already_tried,
  double *coefs
  )
```

In a way, it's kind of silly to pass n_so_far as an input then output. The code just outputs it incremented. But I like being that explicit. This routine also uses npred as shorthand for *n_so_far, a slight simplification.

```
{
  int i, j, k, ivar, ir, npred, n_already_tried, nbest, *root_vars ;
  double crit ;

  *n_so_far = npred = *n_so_far + 1 ; // We just use npred as shorthand for *n_so_far
```

The first step to prepare for adding a new variable is to copy the "best" information to a "prior_best" area from whence it will serve as the basis for all new models. Then reset the "best" area to prepare for this new set of models. We keep a record of the n_already_tried variable sets already tested so that we don't unnecessarily retest combinations.

```
  for (i=0 ; i<nkept ; i++) {
    prior_best_crits[i] = best_crits[i] ;
    memcpy ( &prior_best_trials[i*(npred-1)] ,
         &best_trials[i*(npred-1)] ,
         (npred-1) * sizeof(int) ) ;
    best_crits[i] = -1.e60 ;
    }

  n_already_tried = 0 ; // Counts trial_vars sets tried
```

We will soon pass through all historical best models. The root variable set (best variables from prior step) for each is root_vars. The user may have set nkept huge, in which case we will probably exhaust the best models before hitting this limit. Thus, before looping through the prior best models, we have to figure out how many actual best models there are.

```
  for (ir=0 ; ir<nkept ; ir++) {
    if (prior_best_crits[ir] < -1.e59)    // If we hit the end of prior
      break ;                             // trials, we are done
    }

  nbest = ir ;   // This many best models from last iteration

  for (ir=0 ; ir<nbest ; ir++) {   // Pass through all prior best models
    root_vars = &prior_best_trials[ir*(npred-1)] ;
```

We have a "prior best" set of variables pointed to by root_vars. Try adding each candidate variable. Of course, it makes no sense to add a variable that's already in the root set. So when we consider a new variable ivar, we search the root set for it and skip it if it's already there.

```
for (ivar=0 ; ivar<ncand ; ivar++) {

  // If the trial variable is already in the root set, skip it
  for (i=0 ; i<npred-1 ; i++) {  // ipred has already been incremented, so ipred-1 in root
    if (root_vars[i] == ivar)    // Is this new trial variable already in the root set?
      break ;
    }

  if (i < npred-1)
    continue ; // This trial var is in the root set, so skip it
```

At this point we are inside two loops. The outer of the two gives us a root set root_vars, one of the best variable sets from the prior step. The inner specifies a new variable ivar to append to the root set. These will be merged into the trial_vars array. The straightforward approach would be to just copy the root variables into trial_vars and then append the new trial variable ivar. But it is done very differently here, because we want the variables in trial_vars to always appear in increasing order of their indices (columns in the dataset). This makes redundancy elimination a lot easier, because this ordering ensures that every set of variables has a unique order of appearance in trial_vars. Thus, we loop through the ncand candidate inputs once, in order. For each, first we check the npred-1 root variables to see if it's there. Then we see if it's the current new trial variable ivar.

```
k = 0 ;
for (i=0 ; i<ncand ; i++) {
  for (j=0 ; j<npred-1 ; j++) {
    if (root_vars[j] == i) {  // If this input candidate is in the root list
      trial_vars[k++] = i ;   // Add it to trial_vars for this predictor set
      break ;                 // No need to check further; it can't be there again
      }
    }                         // This loop gets the root variables in the trial
  if (ivar == i)              // Also get this current trial candidate 'ivar'
    trial_vars[k++] = i ;     // We made sure above that ivar is not in root set
  }                           // For building the trial_vars vector
```

At this point, the trial_vars vector contains the (indices of) predictor variables for this trial. There is no sense in retesting predictor sets that have already been tested. So search the array already_tried to see if this set of predictors is there. Do you see why keeping indices sorted is good?

```
for (i=0 ; i<n_already_tried ; i++) {
  for (j=0 ; j<npred ; j++) {
    if (trial_vars[j] != already_tried[i*npred+j])     // Did we find a difference?
      break ;
    }
  if (j == npred)              // If the loop above never broke, we match perfectly
    break ;
  } // Searching the models in already_tried

if (i < n_already_tried)      // If we match, skip this trial set
  continue ;
```

When we get here, we have a trial set that has not been tested. Record it so we don't redo it later, and then compute the R-square performance criterion.

```
for (i=0 ; i<npred ; i++)   // Insert this variable set in array of already tried
  already_tried[n_already_tried*npred+i] = trial_vars[i] ;
++n_already_tried ;

if (xval ( ncases, nfolds, npred, trial_vars, ncand, predvars, target, coefs, &crit ))
  return ERROR_INSUFFICIENT_MEMORY ;
```

We now have in crit the criterion for this trial_vars set. Just as we did when finding the first variable, if it is worthy we insert this set in the array of "best so far" models. This collection is always sorted by performance, with the best at [0] and decreasing from there.

```
for (i=0 ; i<nkept ; i++) {   // See where this trial set fits in the ranking of best so far
  if (crit > best_crits[i])
    break ;                    // It beats the one in slot i
  }

if (i < nkept) {   // We need to move down the existing best and insert this new one

  for (j=nkept-2 ; j>=i ; j--) {
    best_crits[j+1] = best_crits[j] ;
```

```
      memcpy ( &best_trials[(j+1)*npred] , &best_trials[j*npred] , npred * sizeof(int) ) ;
      }

    best_crits[i] = crit ;
    memcpy ( &best_trials[i*npred] , trial_vars , npred * sizeof(int) ) ;
    } // If we are inserting this new 'best' model

  } // For all trial vars
 } // For all root sets
 return 0 ;
}
```

That's it; we've examined every routine related to this superior stepwise selection algorithm. Complete source code is in the file STEPWISE.CPP.

Demonstrating the Algorithm Three Ways

This section presents three examples of the enhanced stepwise selection algorithm. For the first two, the following 11 variables are employed:

> *RAND0–RAND9* are independent (within themselves and with each other) random time series.
>
> *SUM1234 = RAND1 + RAND2 + RAND3 + RAND4*

I specified the minimum and maximum number of variables to be the number of predictor candidates. This forces testing of all candidates. The algorithm produces the following output, slightly reformatted to fit.

```
*****************************************
*                                       *
* Computing enhanced stepwise linear-quadratic model   *
*                                       *
*        SUM1234 is the target          *
*    10  predictor candidates           *
*     5  candidates retained for each iteration   *
*     4  folds for cross validation performance   *
*    10  minimum predictors in final model   *
*                                       *
```

193

```
*    10  maximum predictors in final model            *
*    100  replications of complete Monte-Carlo Test   *
*                                                      *
*******************************************
Stepwise inclusion of variables...
```

R-sqr	MOD pval	CHG pval	Predictors...
0.2811	0.010	0.010	RAND3
0.5183	0.010	0.010	RAND3 RAND4
0.7497	0.010	0.010	RAND2 RAND3 RAND4
1.0000	0.010	0.010	RAND1 RAND2 RAND3 RAND4
1.0000	0.010	0.690	RAND0 RAND1 RAND2 RAND3 RAND4
1.0000	0.010	0.850	RAND0 RAND1 RAND2 RAND3 RAND4 RAND5
1.0000	0.010	0.970	RAND0 RAND1 RAND2 RAND3 RAND4 RAND5 RAND6
1.0000	0.010	1.000	RAND0 RAND1 RAND2 RAND3 RAND4 RAND5 RAND6 RAND7
1.0000	0.010	1.000	RAND0 RAND1 RAND2 RAND3 RAND4 RAND5 RAND6 RAND7 RAND8
1.0000	0.010	1.000	RAND0 RAND1 RAND2 RAND3 RAND4 RAND5 RAND6 RAND7 RAND8 RAND9

```
STEPWISE successfully completed
Final XVAL criterion = 1.00000
In-sample mean squared error = 0.00000
```

Observe the following:

- The R-square criterion jumps up by about 0.25 as each of the four "true" predictors is added, reaching and remaining at 1.0 thereafter.

- Beginning with the first predictor, the model p-value is at the minimum (most significant) possible, 1/mcpt_reps=0.01.

- As the three additional "true" predictors are added, the p-value for the added variable remains at 0.01. But as soon as an irrelevant variable is added, the change p-value jumps up to extreme insignificance. The boundary between important and worthless could not be more clear.

I won't show the results here, but I reran this test with the minimum number of predictors set to 1, the default. It accepted the four "true" predictors exactly as shown previously but stopped with a "performance decrease" caused by addition of a worthless variable.

Finally, here is a more practical example. I computed 19 common indicators used in analyzing equity markets, as well as a measure of the market change over the trading day following availability of the indicators. Here is the output produced by this test:

```
*******************************************
*                                         *
* Computing enhanced stepwise linear-quadratic model  *
*                                         *
*         Z_DAY_RET is the target         *
*    19  predictor candidates             *
*    10  candidates retained for each iteration  *
*     4  folds for cross validation performance  *
*     1  minimum predictors in final model  *
*    19  maximum predictors in final model  *
*   100  replications of complete Monte-Carlo Test  *
*                                         *
*******************************************

Stepwise inclusion of variables...

R-square    MOD pval   CHG pval   Predictors...
  0.0049      0.040      0.040     CMMA_10
  0.0079      0.020      0.090     ADX15 CMMA_10

STEPWISE terminated early because adding a new variable
caused performance degradation

Final XVAL criterion = 0.00793
In-sample mean squared error = 0.98768
```

Regression coefficients for standardized data:

0.035689	ADX15
-0.025106	CMMA_10
-0.001000	ADX15 Squared
0.032440	CMMA_10 Squared
-0.079148	ADX15 * CMMA_10
-0.030700	CONSTANT

The variable first selected, CMMA_10, is the close of the current bar minus the moving average of the prior 10 bars. (All prices are converted to logs before indicator computation is performed.) This variable measures the direction and degree of price departure from recent history. The second variable, ADX15, is the ordinary ADX indicator with a 15-day lookback. This indicates strength of trend, though without specifying a direction.

Even CMMA_10 alone has a p-value of 0.04, meaning that, if CMMA_10 had no day-ahead predictive power, there is only a 0.04 probability that it would have performed as well as it did in predicting market movement the next day. Adding ADX15 lowers this probability to 0.02.

Here's a quick note for mathematically inclined readers. It may superficially appear as if this 0.02 p-value suffers from selection bias and hence is overly optimistic. After all, the program first picked CMMA_10 as the best performer and then picked ADX15 to best complement it. But remember that the permutation replications do exactly the same optimal selections, thus correctly accounting for any selection bias. So this is a fair and unbiased p-value.

The p-value for adding ADX15 is 0.09, decent but not excellent. And after that addition, despite having 17 more industry-standard candidates to choose from, it terminates with the observation that performance deteriorates with the addition of a third indicator.

Finally, I printed the fascinating model coefficients. CMMA_10 has a negative coefficient, alone and in the cross product, which indicates that regression to the mean is at work. And the fact that the cross product has the largest coefficient says that this effect is strongest when it happens in the presence of a strong trend. Very, very interesting!

CHAPTER 6

Nominal-to-Ordinal Conversion

A nominal variable is one that identifies a class membership, as opposed to having a numerical meaning. Nominal variables can have numeric values yet have no numeric meaning, no sense of quantity or order. The classic example is the month of the year. We may say that June has the value 6 and November has the value 11. Certainly 11 is greater than 6, but this does not mean that November is greater than June.

Very few prediction or classification models can directly accept a nominal value as an input, which presents a problem if one or more variables in our application are nominal. There are some awkward ways around this. The most popular method is to recode a single nominal variable as a set of binary variables, with as many binary variables as the nominal variable has classes, and assign one of these binary variables to each class. Then, for each case, we set the single binary variable corresponding to the case's class to 1 and set all others to 0. This works well if there are very few classes, but if there are more than a few classes, not only can this generate an impractical number of input variables, but the information provided by any given class membership can be diluted.

If we have a variable that takes on meaningful numeric values, and that is equal to or shares substantial information with our ultimate target variable, we can often use our training data to elevate the level of an ordinal variable. In theory at least we can elevate it to the same measurement level as our (possibly surrogate) target variable. However, it has been my experience that raising it to just ordinal level, so that order (greater/less than) is meaningful, accomplishes the goal of converting a nominal variable so that it is suitable for model input, without introducing excessive random noise. This will be what is presented in this chapter, although readers should have no trouble revising the code to take the conversion to the level of the "target" if desired.

© Timothy Masters 2020
T. Masters, *Modern Data Mining Algorithms in C++ and CUDA C*,
https://doi.org/10.1007/978-1-4842-5988-7_6

Some nice bells and whistles will be added, but to start, let's discuss the basic idea. The user supplies a dataset containing values of the nominal variable to be converted, as well as values for what we will here call a target variable. In many applications, this will be the actual target variable that will be used by a prediction model. However, all it really needs to be is some variable of at least ordinal level that is significantly related to the ultimate target variable.

As a perhaps overly simplistic example, suppose our ultimate goal is to be able to look at a set of properties of a patient's disease and decide whether a particular treatment should be used as a follow-up to surgery, or if the side effects are too severe to justify it. So this is a binary classification problem: use the treatment or do not use it. Also suppose that we have available as an input variable the ethnic heritage of the patient, perhaps broken down into a dozen or so categories. In an ideal world we would produce 12 different classification models, using a different model for each ethnic category. But the world being what it is, we don't have nearly enough data to take that extravagant approach. So instead we treat ethnicity as a nominal variable and associate it with a synthetic target variable, such as each patient's personal rating of quality of life after treatment, perhaps on a scale of 1 to 10.

We have a training set of patients, all of whom had this treatment. (We may have in our dataset many patients who did not have this treatment. These patients do not concern us now.) In addition to many other measured variables that are not relevant to this discussion, for each patient we have the nominal variable *Ethnicity* and the target variable *Quality of life*. We wish to compute a new variable to substitute for *Ethnicity* that will have a level of measurement higher than nominal so that we can use it as a direct input to our ultimate classification model.

A reasonable approach, which is almost but not quite the approach used here, is to find the mean of the target for each ethnic class and substitute this target mean for the *Ethnicity* variable. For example, suppose people of *Vulcan* ethnicity report a very high quality of life after this treatment, while people of *Romulan* ethnicity report a very low quality of life after the treatment. Then we would recode the dataset, substituting the (large) mean of Vulcans for the *Ethnicity* variable for each Vulcan patient, and the (small) mean of Romulans for each Romulan patient, and similarly for all other ethnicities. This gives us a numeric value for the formerly nominal variable *Ethnicity*, and this new variable can be directly input to a classification or prediction model.

In this particular example, the synthetic target variable is well behaved because it has just ten possible values. But suppose the synthetic target can have heavy tails. For example, perhaps the target is the number of days before death after treatment. Perhaps most patients have about 10–50 days of life remaining, while a very few go on to live many hundreds of days. Taking the mean to use as our substitute value would likely give poor results, because one or a few crazy outliers would skew the results.

To avoid this, in my code I pass through the entire dataset and convert the target values to percentiles. Thus, the case having the smallest target value would have a score of 0, the case with the largest would have a score of 100, and all others would lie in between these extremes. This gives us a new target having ordinal scale; order in the sense of one number being greater than another is preserved, but outliers are tamed. In my own work I have found that this preserves nearly all useful information for the conversion, yet outliers have no impact.

There are three other improvements to this basic technique that I have found to be useful. When I first became involved in predicting movement in financial markets, I soon learned that some techniques work only in times of low volatility and others (fewer!) work only in times of high volatility. I nearly always devised prediction models that specialized in one or the other of these two market states. The same applies to nominal-to-ordinal conversion. It will often be the case that we will want to employ two different conversions, with the choice being dependent on the value of some binary state variable. This binary state variable is often called the *gate*.

A second enhancement to the basic technique is the ability to ignore some cases when devising the nominal-to-ordinal mapping. In some applications we may have reason to believe that the class membership of some cases is irrelevant, with the decision depending on some other variable. Consider the prior example that involved converting the nominal variable *Ethnicity* to an ordinal variable that captures self-assessed quality of life. Suppose that some patients had medical disabilities that prevented them from providing this assessment and a relative substituted his or her own judgement of the patient's quality of life. We may not trust that third-person view and decide, based on a "who answered the question?" variable, whether we want to disregard this case for the purposes of creating the mapping. Of course, we want to avoid creating "missing data" whenever possible, so it is almost always in our best interest to assign some number to such cases. The most reasonable number to assign is the median percentile of the synthetic target, something right in the middle. Naturally, this median will be very close to 50, departing only due to ties in the dataset.

The third desirable enhancement is the ability to decide whether our mapping is based on a legitimate relationship, as opposed to being based on random variation. If the nominal variable that we wish to convert has no legitimate relationship with the synthetic target we are using to compute the mapping, then the whole operation is pointless. We might as well assign random numbers to the cases. This subject is discussed in an upcoming section.

Implementation Overview

The code presented in this chapter and employed in the *VarScreen* program implements gating (being able to compute two separate mappings according to the value of a gate variable) as well as ignoring cases according to a gate value. It handles both options by means of a single gate variable. This does impose some generally minor limitations on the developer. On the other hand, it also simplifies operation of the program. Any reasonably competent programmer should be able to easily modify the supplied code to separate these operations and even to include the possibility of several gate variables, if desired.

This implementation treats the optional gate variable as trinary: positive, negative, or zero. (Any value whose magnitude is less than 1.e-15 is considered to be zero.) A positive value of the gate variable places this case in one mapping category, a negative value places it in another mapping category, and zero causes the case to be ignored. It is legal to use only positive and zero values, or only negative and zero values; either situation will result in only one effective mapping to be generated. And of course, having only positive and negative values means that two maps are produced and no cases are ignored.

Testing for a Legitimate Relationship

We can use a Monte-Carlo permutation test to provide a broadly applicable method for estimating the probability that an apparently decent mapping we obtained could have been nothing more than the product of random, meaningless relationships between our nominal variable and our target. There are many different tests that we could perform. All of them share the null hypothesis that there is no relationship between the nominal variable and the target. But we can test this null hypothesis against a variety of alternative hypotheses. I have chosen the tests shown in the following text. The first test is the only one done if there is no gate variable. If there is a gate variable, three categories are possible for the cases: positive gate mapping, negative gate mapping, and case ignored. If the gate variable takes only positive nonzero values, or only negative nonzero values, the unused "mapping" maps all values of the nominal variable to the median percentile of the target, very close to 50 except in pathological cases of extreme ties. Here are the tests performed:

- The minimum mean target percentile (across all categories of the nominal variable) is subtracted from the maximum. We test whether a difference this extreme could have arisen by random chance from an unrelated nominal variable and target.

- Separately for each nominal variable category, we compute the absolute difference in mean target percentile between the positive gate and the negative gate. We test whether a difference this extreme could have arisen by random luck.

- The maximum of the differences computed previously is considered. We test whether a maximum difference this extreme could be just the product of random luck. The prior test, which looks at each category separately, is subject to selection bias because multiple p-values are computed. This test is immune to this particular selection bias. If you attain a significant p-value in the preceding test, discount its importance if you do not also attain a significant p-value for this test.

- Considering only cases having a negative gate value, compute the minimum mean target percentile across all categories and subtract it from the maximum. We test whether a difference as extreme as what we observed could have arisen by random chance.

- This same test is performed by considering only cases having a positive gate value.

- We look at the greater of the two differences computed in the prior two tests and test whether a maximum difference as large as what we observed could have arisen by random chance. The prior two tests have a small but significant selection bias because we look at each gate category (positive and negative) and pay attention to whichever is more significant. This "greater of the two differences" does not suffer from this particular selection bias.

Of course, in most applications we will be looking at a multitude of p-values, so selection bias is unavoidable. But in order to be able to examine a variety of ways in which the mapping could demonstrate that the null hypothesis (no nominal variable–target relationship) is false, we need to perform multiple tests. Thus, some degree of selection bias is unavoidable.

An Example from Equity Price Changes

The test described in this section is based on over 8500 days of closing prices of OEX, the S&P 100 index. I wondered whether the order of the most recent 3 days' closing prices could be used as an input to a model that predicts price movement the next day. In other words, prices increasing steadily over the prior 3 days might mean one thing, and a steady decrease might mean another, and a price rise followed by a price drop might mean another, and so forth. There are 3!=6 ways in which three different prices can be ordered. (I made arbitrary decisions for ties, which are not terribly common.) Let C be the closing price 2 days ago, B be the closing price yesterday, and A be the closing price today. I assigned the six categories as shown on the next page. To accommodate price ties, the class is assigned to the last category in which it falls.

0:   C <= B <= A
1:   C <= A <= B
2:   B <= C <= A
3:   B <= A <= C
4:   A <= B <= C
5:   A <= C <= B

This is clearly a nominal variable, as there is no apparent way to sensibly assign numeric values to these categories.

It is well known that market price patterns can take very different forms in times of high volatility as opposed to times of low volatility. I decided that I wanted to compute separate mappings for exceptionally high volatility regimes and exceptionally low volatility regimes and ignore price order when volatility is just average. (This may or may not be a wise plan in real life, but it ideally suits demonstrating this mapping technique.) This test produced the following output:

```
*********************************************************************
*                                                                   *
* Computing nominal-to-ordinal conversion                           *
*                                                                   *
*        ORDER_CLASS is the sole predictor                          *
*             VOLATILE is the gate                                  *
*             Z_DAY_RET is the target                               *
*      1000 replications of complete Monte-Carlo Test               *
*                                                                   *
*********************************************************************
```

Class bin counts...

Class	Gate-	Gate0	Gate+
0	874	707	710
1	471	324	378
2	473	325	360
3	380	324	350
4	687	552	581
5	412	327	313

Class bin mean percentiles...

Class	Gate-	Gate0	Gate+
0	47.40	50.09	48.13
1	48.77	49.68	51.97
2	48.04	46.74	50.96
3	50.03	52.56	50.26
4	50.87	50.39	54.61
5	51.73	49.73	50.34

For each class individually, p-value for positive gate versus negative gate...

Class	p-value
0	0.614
1	0.097
2	0.145

3	0.939
4	0.018
5	0.532

p-value for max across classes of the gate +/- difference = 0.254
p-value for max class mean percentile minus min, for negative gate   = 0.162
p-value for max class mean percentile minus min, for positive gate   = 0.016
p-value for max of the above two = 0.024

By examining the bin counts, we see (not surprisingly) that for all volatility regimes, the pattern of steadily increasing price is by far the most common. A fairly distant second is the pattern of steadily decreasing prices.

The target is the log price change the next day. The table of mean target percentiles shows an interesting pattern. For both extremes of volatility, the category of steadily increasing prices shows the smallest mean target percentile. For exceptionally high volatility, the category of steadily decreasing prices shows the largest mean target percentile, and for exceptionally low volatility, this category is second largest. The largest is still a category in which the most recent price is the lowest of the three. This is evidence that mean reversion is in control, as opposed to trend following, at least for these two extremes of volatility.

Note that the "Gate 0" category, which means "ignore this case," still has mean target percentiles printed for the edification of the user. When the new nominal variable is generated, it will be assigned the target median percentile, which of course will be very close to 50.

Now let's look at the p-values. We see that the difference in target mean percentiles is highly significant (0.018) for only category 4, steadily decreasing prices. However, we are picking the most significant out of six p-values, so selection bias is at work. We see that when the max across categories is considered, the p-value is an unimpressive 0.254. This tells us that we should not pay much attention to that one attention-grabbing p-value. It could easily be the product of random variation.

The maximum difference across classes for a negative gate (exceptionally low volatility) is an uninteresting 0.162. But with high volatility, we get a much more impressive p-value of 0.016. Moreover, we are inclined to take it seriously, since the selection-bias-free p-value for categories is 0.024, quite respectable.

After this test is run, a new variable is created. If this is the first time during this run of the *VarScreen* program that we performed nominal-to-ordinal conversion, this variable will be named NomOrd_1. The second time, the new variable will be NomOrd_2 and so forth. This variable can be used in subsequent tests, and it can also be written to a text file using the "File / Write variables" menu command.

Code for Nominal-to-Ordinal Conversion

In order to use this code, you will need to do the following:

1) Create a new NomOrd object by invoking the constructor with the following parameters:

   ncases – Number of cases in training set

   npred – Number of predictors (1 if class, >1 if max)

   preds – ncases by npred matrix of predictors

   gate – ncases vector of gate variable; NULL if no gate

   The ***npred*** parameter and ***preds*** array/matrix require a little explanation. If ***npred***=1 then ***preds*** is an array of class IDs. Every element of this array will be rounded to the nearest integer (not truncated) to get the class ID for that case. IDs must begin at 0. If any ID is negative, it will be treated as if it is 0. It is strongly suggested (though not required) that no ID between 0 and the largest be skipped. If ***npred***>1 then it is the number of classes and, for each case, whichever column of the ***preds*** matrix has the largest entry determines the class membership of that case. Ties may be broken randomly.

2) If desired, call the print_counts (pred_index) member function. The pred_index parameter is used only to allow printing variable names. See that code for details, and know that you can omit it or revise the naming process.

3) Call the train (target) member function, passing it a target vector. This function may be called as often as desired with different target vectors. This function computes the mapping that maps classes (categories) of the nominal variable **preds** to mean target percentiles (percentiles).

4) If desired, call the print_ranks (pred_index) member function. This prints the mapping as a table of mean percentiles for each category.

5) If desired, call the mcpt(mcpt_type , mcpt_reps , target , pred_index) member function. This computes and prints the p-values described in the prior section.

The Constructor

Here is the calling parameter list and some memory allocations. More memory will be allocated soon. For clarity, error checking has been omitted. See NOM_ORD.CPP for full details.

```
NomOrd::NomOrd (
  int ncases_in ,            // Number of cases in training set
  int npred_in ,             // Number of predictors (1 if class, >1 if max)
  double *preds_in ,         // ncases_in by npred_in matrix of predictors
  double *gate_in            // ncases vector of gate variable; NULL if no gate
  )
{
  int i, j, jbig ;
  double biggest ;

  ncases = ncases_in ;       // Save private copies in the object
  npred = npred_in ;

  class_id = (int *) MALLOC ( ncases * sizeof(int) ) ;
  if (gate_in != NULL)       // We'll convert user's real gate into -1 / 0 / +1 trinary flag
    gate = (int *) MALLOC ( ncases * sizeof(int) ) ;
  else
    gate = NULL ;
```

```
ranks = (double *) MALLOC ( ncases * sizeof(double) ) ;
indices = (int *) MALLOC ( ncases * sizeof(int) ) ;
temp_target = (double *) MALLOC ( 2 * ncases * sizeof(double) ) ;
target_work = temp_target + ncases ;
```

The next loop passes through all cases. Integer class IDs are placed in class_id, and if
the user supplies a gate array, trinary integer gate flags are placed in gate.

```
nclasses = npred ;              // If npred==1 this will keep track of largest ID so far
for (i=0 ; i<ncases ; i++) {
  if (npred == 1) { // The class is the single predictor, rounded, origin 0
    class_id[i] = (int) (preds_in[i] + 0.5) ;
    if (class_id[i] < 0)
      class_id[i] = 0 ;
    if (class_id[i]+1 > nclasses) // Class ID is origin 0
      nclasses = class_id[i]+1 ;
    }
  else {          // The class is whichever predictor has largest value
    for (j=0 ; j<npred ; j++) {
      if (j == 0 || preds_in[i*npred+j] > biggest) {
        biggest = preds_in[i*npred+j] ;
        jbig = j ;
        }
      }
    class_id[i] = jbig ;
    }

  if (gate != NULL) {
    if (gate_in[i] > 1.e-15)
      gate[i] = 1 ;
    else if (gate_in[i] < -1.e-15)
      gate[i] = -1 ;
    else
      gate[i] = 0 ;
    }
  }
```

The first block of code in the preceding ncases loop handles the case of npred=1. Each value in preds_in is the actual class ID. We round to the nearest integer, bound it at 0 to prevent negative IDs, and keep track of the largest class ID.

The second block of code handles the case of having a separate variable for each possible class. For each case, we see which variable has the largest value and set the class ID accordingly.

The final block converts the real-valued gate variable to -1, 0, or +1 according to the sign and magnitude of the original gate. This speeds processing later.

Now that we know the number of classes (categories for the nominal variable), we can finish memory allocation as shown on the next page. If there is a gate, we need more allocations than if there is no gate.

The items that we allocate have the following uses:

> class_counts – This counts the number of cases in each class, considering any gate for counting or ignoring a case. If there is no gate this will be used for map computation. If there is a gate it will still be computed but never used. (It might be useful to print this for the user or find some other use.)

> mean_ranks – The mapping function is computed and saved here. It will be printed for the user to allow the user to write a transformation function for their application. This will also provide the data for the Monte-Carlo permutation test.

> bin_counts – This is the gated version of class_counts. If there is no gate, this is not used. If there is a gate, this counts the number of cases in each bin, where a bin is defined by its nominal variable category (class) and its trinary gate value.

> orig_gate – This stores the unpermuted criterion vector for one of the Monte-Carlo permutation tests.

> count_gate – This is the counter for that test. All other MCPT original values and counters are scalars and hence do not need memory allocation.

```
class_counts = (int *) MALLOC ( nclasses * sizeof(int) ) ;

if (gate == NULL) {
  mean_ranks = (double *) MALLOC ( nclasses * sizeof(double) ) ;
  bin_counts = NULL ;
  orig_gate = NULL ;
  count_gate = NULL ;
  }

else {
  mean_ranks = (double *) MALLOC ( 3 * nclasses * sizeof(double) ) ;
  bin_counts = (int *) MALLOC ( 3 * nclasses * sizeof(int) ) ;
  orig_gate = (double *) MALLOC ( nclasses * sizeof(double) ) ;
  count_gate = (int *) MALLOC ( nclasses * sizeof(int) ) ;
  }
```

Several items relevant to this discussion are declared in the constructor but do not
need allocation:

```
int gate_counts[3] ;       // Number of cases in each gate bin
double orig_class[2] ;     // Rep 0 max class difference for gate- and gate+
int count_class[2] ;       // Counter for this permutation test
```

The final step is to count the number of cases in each bin. The first block of code in the
following zeros all counters. Then we loop through the cases to count cases in each bin.

```
for (i=0 ; i<nclasses ; i++) {
  class_counts[i] = 0 ;
  if (gate != NULL)
    bin_counts[i*3+0] = bin_counts[i*3+1] = bin_counts[i*3+2] = 0 ;
  }

gate_counts[0] = gate_counts[1] = gate_counts[2] = 0 ; // Negative, zero, positive

for (i=0 ; i<ncases ; i++) {
  if (gate == NULL)
    ++class_counts[class_id[i]] ;

  else {
    if (gate[i])
      ++class_counts[class_id[i]] ;
```

```
     if (gate[i] < 0) {
       ++gate_counts[0] ;
       ++bin_counts[class_id[i]*3+0] ;
       }
     else if (gate[i] > 0) {
       ++gate_counts[2] ;
       ++bin_counts[class_id[i]*3+2] ;
       }
     else {
       ++gate_counts[1] ;
       ++bin_counts[class_id[i]*3+1] ;
       }
     }
   }
}
```

Printing the Table of Counts

Most readers will want to print counts in their own preferred format. However, the printing code used in *VarScreen* is shown here simply because it clarifies the use and layout of the bin count arrays. This information is available for printing as soon as the constructor is created.

```
void NomOrd::print_counts ( int *pred_index )
{
   int i ;
   char msg[256] ;

   if (gate == NULL) {  // No gate is being used
     if (npred == 1) {    // The single nominal variable contains class IDs
       audit ( " Class Count" ) ;  // Thus, we do not have variable names for the classes
       for (i=0 ; i<nclasses ; i++) {    // Class IDs are just numbers, not names
         sprintf_s ( msg, "%6d %8d", i, class_counts[i] ) ;
         audit ( msg ) ;
         }
       }
```

```
    else {      // We have a separate variable for each class; we can name the classes
      audit ( "    Predictor Count" ) ;
      for (i=0 ; i<npred ; i++) {
        sprintf_s ( msg, "%15s %8d", var_names[pred_index[i]], class_counts[i] ) ;
        audit ( msg ) ;
        }
      }
    } // If gate == NULL

  else {      // A gate is being used
    if (npred == 1) {   // The single nominal variable contains class IDs
      audit ( "Class Gate- Gate0 Gate+" ) ;
      for (i=0 ; i<nclasses ; i++) {
        sprintf_s ( msg, "%5d %8d %8d %8d",
                    i, bin_counts[i*3+0], bin_counts[i*3+1], bin_counts[i*3+2] ) ;
        audit ( msg ) ;
        }
      }

    else {      // We have a separate variable for each class; we can name the classes
      audit ( "    Predictor Gate- Gate0 Gate+" ) ;
      for (i=0 ; i<npred ; i++) {
        sprintf_s ( msg, "%15s %8d %8d %8d", var_names[pred_index[i]],
                    bin_counts[i*3+0], bin_counts[i*3+1], bin_counts[i*3+2] ) ;
        audit ( msg ) ;
        }
      }
    }
  }
}
```

The only thing to take note of in this code is the use of the mysterious pred_index parameter. Readers may take a different approach. In *VarScreen*, there is a master database of all variables, including those not being used in the nominal-to-ordinal study. The names of these variables are stored in the global var_names, so the caller passes in pred_index the indices in the master database of the variables being used in this particular study. This lets us print the names of the variables. Naturally, each reader would want to print variable names in a way that is appropriate for his or her code.

It should also be clear that the function audit() just prints to a log file whatever line of characters is passed to it. Readers may, of course, also wish to customize this printing method.

Computing the Mapping Function

To compute the function that maps the previously provided nominal variable and gate to a target, the user calls train(). This may be done as often as desired, with different target vectors. The first step is to convert the target to ranks (actually, percentiles), including ties. Simultaneously get the grand median percentile, which we need later. Then sort again using the indices to put them back in order so that they correspond to the predictors. Of course we could do the unsorting faster using a work array, but this method is not much slower and a lot simpler (one line of code!).

```
void NomOrd::train (
  double *target )    // Target variable, not disturbed
{
  int i, k ;
  double val, rank ;

  for (i=0 ; i<ncases ; i++) {
    ranks[i] = target[i] ;      // We must not disturb the caller's target
    indices[i] = i ;            // Keeps track of original indices after sorting
    }

  qsortdsi ( 0 , ncases-1 , ranks , indices ) ; // Sort ascending, moving indices


  for (i=0 ; i<ncases ; ) {   // Convert target values to ranks
    val = ranks[i] ;
    for (k=i+1 ; k<ncases ; k++) { // Find all ties
      if (ranks[k] > val)
        break ;
      }
```

```
  rank = 0.5 * ((double) i + (double) k + 1.0) ;
  while (i < k)
     ranks[i++] = 100.0 * rank / ncases ; // Convert rank to percentile
     } // For each case in sorted y array

if (ncases % 2)
  median = ranks[ncases/2] ;
else
  median = 0.5 * (ranks[ncases/2-1] + ranks[ncases/2]) ;

qsortisd ( 0 , ncases-1 , indices , ranks ) ;    // Undo the sorting to restore original order
```

Now we pass through the data, cumulating the mean percentile for each bin. If there is no gate, we need one bin in mean_ranks for each class. But if there is a gate, we need three bins for each class, because positive, negative, and zero gates will each be tallied separately. Recall that we counted the number of cases in each bin when we executed the constructor. So here all we have to do is sum the ranks that fall in each bin. The first block of code in the following, which zeros all summation bins, shows how the classes and gate values are arranged in the bin array.

```
for (i=0 ; i<nclasses ; i++) {
  if (gate == NULL)
    mean_ranks[i] = 0 ;
  else
    mean_ranks[i*3+0] = mean_ranks[i*3+1] = mean_ranks[i*3+2] = 0.0 ;
  }
```

The next code block handles the case of there being no gate. It passes through all cases, summing the bins, and then divides by the previously computed bin counts to get the mean for each bin. In the unusual situation that a bin has no cases, we set its "mean rank" to the grand median rank (percentile).

```
if (gate == NULL) {
  for (i=0 ; i<ncases ; i++) {
    k = class_id[i] ;        // Class of this case
    mean_ranks[k] += ranks[i] ; // Cumulate target rank of this case
    }
```

```
for (i=0 ; i<nclasses ; i++) {
  if (class_counts[i] > 0)
    mean_ranks[i] /= class_counts[i] ;
  else
    mean_ranks[i] = median ;
  }
}
```

The code shown on the next page handles the case of having a gate. It's just a simple generalization of the preceding code, checking the gate sign and assigning each case to the appropriate bin according to its class and gate value.

```
else {
  for (i=0 ; i<ncases ; i++) {
    k = class_id[i] ;        // Class of this case
    if (gate[i] < 0)
      mean_ranks[k*3+0] += ranks[i] ;   // Negative gate
    else if (gate[i] > 0)
      mean_ranks[k*3+2] += ranks[i] ;   // Positive gate
    else
      mean_ranks[k*3+1] += ranks[i] ;   // Zero gate (ignore this case for mapping)
  }
```

Bin ranks are summed for this situation of having a gate. Now divide by the case count in each bin to get the mean ranks. If a bin has no cases, we use the grand median rank as its "mean rank."

```
for (i=0 ; i<nclasses ; i++) {

  if (bin_counts[i*3+0] > 0)     // Negative gate
    mean_ranks[i*3+0] /= bin_counts[i*3+0] ;
  else
    mean_ranks[i*3+0] = median ;

  if (bin_counts[i*3+1] > 0)     // Zero gate (ignore this case for mapping)
    mean_ranks[i*3+1] /= bin_counts[i*3+1] ;
  else
    mean_ranks[i*3+1] = median ;
```

```
      if (bin_counts[i*3+2] > 0)      // Positive gate
        mean_ranks[i*3+2] /= bin_counts[i*3+2] ;
      else
        mean_ranks[i*3+2] = median ;
      }
    }

}
```

Monte-Carlo Permutation Tests

The routine that performs the Monte-Carlo permutation tests is the longest and most complicated of all of these routines; keeping track of original and permuted values of all key mapping computations is messy. Here is the beginning of this routine. Note that pred_index has nothing to do with this algorithm; it is used solely for printing class names. See the discussion of printing the table of counts on Page 210 for details on the meaning and use of this otherwise unnecessary calling parameter. As usual, the first replication is unpermuted; subsequent reps permute the target.

```
void NomOrd::mcpt (
   int type ,           // Type of shuffling: 1=complete, 2=cyclic
   int reps ,           // Number of replications
   double *target ,     // Target variable, not disturbed
   int *pred_index      // Database variable indices for printing only
   )
{
   int i, j, irep ;
   double dtemp, min_neg, max_neg, min_pos, max_pos, max_gate ;
   char msg[256] ;

   if (reps < 1) // It would be silly to call this with nreps <= 1
      reps = 1 ; // And even sillier to call it with zero

   for (irep=0 ; irep<reps ; irep++) { // All replications (including the first, unpermuted)
      memcpy ( temp_target , target , ncases * sizeof(double) ) ; // Don't damage target
```

```
if (irep) { // Replications past the first permute the target.

   if (type == 1) {      // Complete permutation (best but illegal if serial correlation)
      i = ncases ;        // Number remaining to be shuffled
      while (i > 1) {      // While at least 2 left to shuffle
        j = (int) (unifrand() * i) ;
        if (j >= i)
          j = i - 1 ;
        dtemp = temp_target[--i] ;
        temp_target[i] = temp_target[j] ;
        temp_target[j] = dtemp ;
        }
      } // Type 1, Complete
   else if (type == 2) { // Cyclic permutation is required if target has serial correlation
      j = (int) (unifrand() * ncases) ;
      if (j >= ncases)
      j = ncases - 1 ;
      for (i=0 ; i<ncases ; i++)
        target_work[i] = temp_target[(i+j)%ncases] ;
      for (i=0 ; i<ncases ; i++)
        temp_target[i] = target_work[i] ;
      } // Type 2, Cyclic
   } // If in permutation run (irep > 0)
```

As has been discussed several times before, if at least one predictor *and* the target have serial correlation, we do not dare use the otherwise superior complete permutation algorithm. Instead we must "permute" by rotating the target with endpoint wraparound. This provides less thorough permutation than the complete algorithm, but it does mostly preserve serial dependencies, resulting in a more accurate p-value than would be obtained with complete permutation.

We now call the train() routine to compute mean ranks. If this is the first, unpermuted replication, we save the original value of every test statistic and we initialize all counters to 1.

```
train ( temp_target ) ;

if (irep == 0) {
  if (gate == NULL) {
    for (i=0 ; i<nclasses ; i++) {
      if (i == 0)
        min_pos = max_pos = mean_ranks[i] ;
      else {
        if (mean_ranks[i] > max_pos)
          max_pos = mean_ranks[i] ;
        if (mean_ranks[i] < min_pos)
          min_pos = mean_ranks[i] ;
        }
      } // For i<nclasses
    orig_max_class = max_pos - min_pos ;
    count_max_class = 1 ;
    } // gate == NULL
```

The preceding code handles the relatively simple case of having no gate. In this case, there is only one test statistic, the difference between the maximum mean rank across all classes and the minimum. So it just computes these quantities and saves this original value in orig_max_class.

When there is a gate, there are multiple test statistics. Separately for each class we save in orig_gate[i] the absolute difference between the positive and negative gate mean ranks. The maximum of these values goes in orig_max_gate. And separately for the positive and negative gates we find the maximum and minimum mean ranks across all classes. The max minus min for negative gates goes in orig_class[0] and that for positive gates goes in orig_class[1]. Finally, the max of these two goes in orig_max_class.

```
else {
  for (i=0 ; i<nclasses ; i++) {
    orig_gate[i] = fabs ( mean_ranks[i*3+0] - mean_ranks[i*3+2] ) ;
    count_gate[i] = 1 ;
    if (i == 0) {
      orig_max_gate = orig_gate[i] ;
      min_neg = max_neg = mean_ranks[i*3+0] ;
```

```
            min_pos = max_pos = mean_ranks[i*3+2] ;
            }
        else {
          if (orig_gate[i] > orig_max_gate)
            orig_max_gate = orig_gate[i] ;
          if (mean_ranks[i*3+0] > max_neg)
            max_neg = mean_ranks[i*3+0] ;
          if (mean_ranks[i*3+0] < min_neg)
            min_neg = mean_ranks[i*3+0] ;
          if (mean_ranks[i*3+2] > max_pos)
            max_pos = mean_ranks[i*3+2] ;
          if (mean_ranks[i*3+2] < min_pos)
            min_pos = mean_ranks[i*3+2] ;
            }
        } // For i<nclasses
      orig_class[0] = max_neg - min_neg ;
      orig_class[1] = max_pos - min_pos ;
      orig_max_class = (orig_class[0] > orig_class[1]) ? orig_class[0] : orig_class[1] ;
      count_max_class = count_max_gate = count_class[0] = count_class[1] = 1 ;
      } // gate not NULL
    } // If irep==0
```

If we are past the first replication, and hence permuted, then we compute the same quantities as those computed in the prior code. But now instead of saving them, we compare them to their original unpermuted values. If the permuted value equals or exceeds the original value, we increment the corresponding counter.

Again, things are simplest when there is no gate. The only test statistic is the difference between the maximum mean rank and the minimum.

```
  else { // We are in a permutation rep
    if (gate == NULL) {
      for (i=0 ; i<nclasses ; i++) {
        if (i == 0)
          min_pos = max_pos = mean_ranks[i] ;
        else {
```

```
        if (mean_ranks[i] > max_pos)
          max_pos = mean_ranks[i] ;
        if (mean_ranks[i] < min_pos)
          min_pos = mean_ranks[i] ;
        }
      } // For i<nclasses
   if (max_pos - min_pos >= orig_max_class)
      ++count_max_class ;
   } // gate == NULL
```

When there is a gate, we compute multiple statistics, exactly as before. I won't waste time by explaining this voluminous code line by line. It corresponds exactly to what we saw in the unpermuted case, except that now instead of saving the values, we compare them to the original values and increment the counters accordingly.

```
else {
  for (i=0 ; i<nclasses ; i++) {
    if (fabs ( mean_ranks[i*3+0] - mean_ranks[i*3+2] ) >= orig_gate[i])
      ++count_gate[i] ;
    if (i == 0) {
      max_gate = fabs ( mean_ranks[i*3+0] - mean_ranks[i*3+2] ) ;
      min_neg = max_neg = mean_ranks[i*3+0] ;
      min_pos = max_pos = mean_ranks[i*3+2] ;
      }
    else {
      if (fabs ( mean_ranks[i*3+0] - mean_ranks[i*3+2] ) > max_gate)
        max_gate = fabs ( mean_ranks[i*3+0] - mean_ranks[i*3+2] ) ;
      if (mean_ranks[i*3+0] > max_neg)
        max_neg = mean_ranks[i*3+0] ;
      if (mean_ranks[i*3+0] < min_neg)
        min_neg = mean_ranks[i*3+0] ;
      if (mean_ranks[i*3+2] > max_pos)
        max_pos = mean_ranks[i*3+2] ;
      if (mean_ranks[i*3+2] < min_pos)
        min_pos = mean_ranks[i*3+2] ;
      }
    } // For i<nclasses
```

```
      if (max_gate >= orig_max_gate)
        ++count_max_gate ;
      if (max_neg - min_neg >= orig_class[0])
        ++count_class[0] ;
      if (max_pos - min_pos >= orig_class[1])
        ++count_class[1] ;
      if ((((max_neg-min_neg) > (max_pos-min_pos)) ?
          (max_neg-min_neg) : (max_pos-min_pos)) >= orig_max_class)
        ++count_max_class ;
      } // gate not NULL
    } // If irep==0
  } // For irep
```

All that remains is to print the results. You will likely want to print the results your own way or just save them for printing later. But my code may clarify the meaning of the quantities just computed. On Page 200 you will find a more detailed explanation of the meaning of each of the p-values computed and printed here. Also see Page 210 for a discussion of how the otherwise unused variable pred_index is used here.

```
if (gate == NULL) {
  sprintf_s ( msg, "p-value for max mean rank minus min mean rank = %.3lf",
              (double) count_max_class / (double) reps ) ;
  audit ( msg ) ;
  }

else {
  audit( "For each class individually, p-value for positive gate versus negative gate...");
  if (npred == 1) { // A single variable specifies the class ID as an integer
    audit ( " Class p-value" ) ;
    for (i=0 ; i<nclasses ; i++) {
      sprintf_s ( msg, "%6d %8.3lf", i, (double) count_gate[i] / (double) reps ) ;
      audit ( msg ) ;
      }
    }
```

```
  else {      // We have a separate variable for each class
    audit ( "    Predictor p-value" ) ;
    for (i=0 ; i<npred ; i++) { // Recall that npred == nclasses
      sprintf_s ( msg, "%15s %8.3lf",
                  var_names[pred_index[i]], (double) count_gate[i] / (double) reps ) ;
      audit ( msg ) ;
      }
    }

  audit ( "" ) ;
  sprintf_s ( msg, "p-value for max across classes of the gate +/- difference = %.3lf",
        (double) count_max_gate / (double) reps ) ;
  audit ( msg ) ;

  sprintf_s ( msg,
          "p-value for max class mean rank minus min, for negative gate = %.3lf",
          (double) count_class[0] / (double) reps ) ;
  audit ( msg ) ;

  sprintf_s ( msg,
          "p-value for max class mean rank minus min, for positive gate = %.3lf",
          (double) count_class[1] / (double) reps ) ;
  audit ( msg ) ;

  sprintf_s ( msg, "p-value for max of the above two = %.3lf",
          (double) count_max_class / (double) reps ) ;
  audit ( msg ) ;
  } // Gate is not NULL

}
```

Index

© Timothy Masters 2020
T. Masters, *Modern Data Mining Algorithms in C++ and CUDA C*,
https://doi.org/10.1007/978-1-4842-5988-7

Printed in the United States
By Bookmasters